THE MORNING
after
EARTH DAY

THE MORNING
after
EARTH DAY

Practical
Environmental
Politics

MARY GRAHAM

GOVERNANCE INSTITUTE
BROOKINGS INSTITUTION PRESS
Washington, D.C.

Copyright © 1999
THE BROOKINGS INSTITUTION
1775 Massachusetts Avenue, N.W., Washington, D.C. 20036
www.brookings.edu
All rights reserved

Library of Congress Cataloging-in-Publication data

Graham, Mary (Mary W.), 1944–
    The morning after earth day : practical environmental
politics / Mary Graham.
        p.   cm.
    Includes index.
    ISBN 0-8157-3236-8 (alk. paper)
    ISBN 0-8157-3235-X (pbk : alk. paper)
    1. Environmental policy—United States. 2. Environmental
protection—United States. I. Title.
GE180 .G73 1999   98-58141
363.7′05′0973—dc21   CIP

Typeset in Palatino

Composition by AlphaWebTech
Mechanicsville, Maryland

Printed by R. R. Donnelley and Sons
Harrisonburg, Virginia

THE BROOKINGS INSTITUTION

The Brookings Institution is an independent, nonprofit organization devoted to nonpartisan research, education, and publication in economics, government, foreign policy, and the social sciences generally. Its principal purposes are to aid in the development of sound public policies and to promote public understanding of issues of national importance. The Institution was founded on December 8, 1927, to merge the activities of the Institute for Government Research, founded in 1916, the Institute of Economics, founded in 1922, and the Robert Brookings Graduate School of Economics and Government, founded in 1924.

The Institution maintains a position of neutrality on issues of public policy to safeguard the intellectual freedom of the staff. Interpretations or conclusions in Brookings publications should be understood to be solely those of the authors.

THE GOVERNANCE INSTITUTE

The Governance Institute, a nonprofit organization incorporated in 1986, is concerned with exploring, explaining, and easing problems associated with both the separation and division of powers in the American federal system. It is interested in how the levels and branches of government can best work with one another. It is attentive to problems within an organization or between institutions that frustrate the functioning of government. The Governance Institute is concerned as well with those professions and mediating groups that significantly affect the delivery and quality of public services. The Institute's focus is on institutional process, a nexus linking law, institutions, and policy. The Institute believes that problem solving should integrate research and discussion. This is why the Institute endeavors to work with those decisionmakers who play a role in making changes in process and policy. The Institute currently has four program areas: problems of the judiciary; problems of the administrative state; problems in criminal justice; and challenges to the legal profession.

*For my husband Don and
my children Liza, Laura, Will, and Molly*

Foreword

AS WE APPROACH the thirtieth anniversary of Earth Day, congressional debate about pollution control and conservation often remains doctrinaire. But across the country, environmental pragmatism is gaining ground. Policymakers, business executives, and citizen groups are fighting novel battles and sometimes making peace with surprising compromises.

After a generation of progress in reducing large sources of industrial and municipal pollution and in improving management of public lands, however, today's environmental conflicts are more complex. They involve controlling pollution caused by farmers, small businesses, drivers of aging cars, and homeowners, as well as minimizing ecological threats on private land. Remedies often lie in politically treacherous territory—persuading ordinary people to change their daily routines rather than ordering big business to adopt new technology or government officials to manage public lands differently.

In this copublication of the Brookings Institution and the Governance Institute, Mary Graham examines the ways in which practical approaches to these conflicts are providing clues for future policy. She also explores core dilemmas that remain, including how to reconcile environmental protection with respect for private property, how to balance federal and state authority, and how much to rely on changing habits as opposed to changing technologies.

Mary Graham is a research fellow at the A. Alfred Taubman Center for State and Local Government and the Environment and Natural Resources Program of Harvard University's Kennedy School of Government and a fellow of the Governance Institute. She is a lawyer who has written about public policy issues for *The Atlantic Monthly, The New Republic*, and other publications.

At Brookings, the author is grateful to the Governmental Studies program directed by Thomas E. Mann for its support of this project. John D. Donahue, J. Clarence Davies, and Barry Rabe provided detailed and helpful comments to Brookings as outside reviewers. Vicky Macintyre edited the manuscript and Carole Plowfield verified factual information. Susan Woollen provided expertise in the production process. Kristen Lippert-Martin provided administrative support. Julia Petrakis prepared the index and Carlotta Ribar proofread final copy.

The views expressed here are those of the author and should not be ascribed to people or organizations acknowledged above, to the trustees, officers, and staff members of the Brookings Institution, or to the directors, officers, or other staff members of the Governance Institute.

MICHAEL H. ARMACOST
President

Washington, D.C.
May 1999

Preface

THIS BOOK BEGAN as an effort to solve a puzzle. While researching an article for the *Atlantic Monthly*, I was struck by an anachronism. National debate about environmental policy remained polarized, but around the country issues were being resolved in ways that seemed strikingly pragmatic. Viewed from Washington, the impression was one of stalemate. Viewed from elsewhere, the impression was one of remarkable adaptation to changing problems and changing times.

As I read and interviewed more widely, I became convinced that the key to this puzzle lay in the fact that the nation's environmental policy was in the midst of a fundamental transition. Assumptions that formed the foundation for laws in the 1970s were being replaced by new ideas. National debate, however, was lagging behind. Because patterns of change were not yet widely recognized, I decided to try to capture the character of this transition in a long essay. The chapters that follow are the result of that effort. They assess the evolving nature of environmental problems addressed by national and state policy, explore changing assumptions, and describe a new style of environmental politics. During the two years that I have spent trying to understand these issues, I have incurred many debts.

I am grateful to the Kennedy School of Government at Harvard University for providing an ideal location for this work. Alan Altshuler, Ruth and Frank Stanton Professor in Urban Policy and Planning and director of the A. Alfred Taubman Center for State and Local Government, and Henry Lee, director of the Environment and Natural Resources Program, offered me a joint appointment as a research fellow and provided invaluable insights at every stage of the project. Robert N. Stavins, Albert Pratt Professor of Business and Government, was unfailingly generous with his time and advice. Both in conversation and through his writings, William C. Clark, Harvey Brooks Professor of International Science, Public Policy, and Human Development, helped me understand the complexities of decisionmaking based on incomplete information. Arnold Howitt, executive director of the Taubman Center, and David Luberoff, associate director, shared their extensive knowledge of environmental issues. Thomas M. Parris, environmental resources librarian at the Harvard college library, used his remarkable grasp of information to provide frequent advice about complex research questions.

This work could not have been undertaken without the support of the Governance Institute and its board of directors. The Institute was the moving force behind the project and Robert A. Katzmann, the Institute's president and acting director of the Brookings Governmental Studies program during the project's formative stages, provided constant encouragement and advice.

I owe a special debt to Paul Portney, president of Resources for the Future, whose efforts to help improve the accuracy and clarity of this work have been extraordinary. I am grateful to historian Michael R. Beschloss for insights into the difficult task of writing about recent history, of which he is a master. Lois Schiffer, assistant attorney general of the Environment and Natural Resources Division of the U.S. Department of Justice, taught me to appreciate the complexities of legal issues involved in federal policy. George T. Frampton Jr., now acting chairman of the President's Council on Environmental Quality, offered many insights drawn from his broad experience. Finally, I will never be able to repay my debt to Bob Samuelson, my close friend and valued critic since we were both teenagers, who provided fresh ideas and asked hard questions at every stage of this work.

I am also grateful to the many people who read earlier drafts of the book. In addition to those mentioned above, I want to thank Senator

Daniel Patrick Moynihan and James P. Leape, vice president of the World Wildlife Fund, for their helpful comments. John Cronan, Jim Scafide, and Francesca Bosco contributed valuable research assistance. Any errors, of course, are the responsibility of the author, and not of those who have worked to minimize them.

Finally, I want to thank my parents, Dr. Robert William Wissler and Elizabeth Anne Wissler, a professor of medicine and a social worker by profession and lifelong naturalists by inclination. They raised their four children on Chicago's south side where both the power and pollution of the steel mills and the ever-changing beauty of Lake Michigan were part of daily life.

Contents

Introduction

ON WEDNESDAY, APRIL 22, 1970, a mild spring day in Manhattan, police closed streets surrounding the city's bustling Union Square at noon and kept them closed until midnight. A hundred thousand office workers, neighborhood residents, and visitors streamed onto the impromptu pedestrian mall, stopping at one or another of the 100 booths set up to promote a variety of causes—urban planning, conservation of natural resources, voluntary sterilization—or stepping inside a polyethylene bubble to take a few breaths of pure, filtered air. Crowds listened to speakers ranging from Mayor John V. Lindsay to entertainer Arthur Godfrey and anthropologist Margaret Mead. At a giant 5 p.m. rally, they were entertained by Leonard Bernstein, Paul Newman, Dustin Hoffman, Pete Seeger, and the cast of the musical *Hair.* Uptown on Fifth Avenue, closed to traffic from noon until 2 p.m., thousands more strolled and picnicked, entertained by singers and fashion shows.

The occasion for this rare hiatus in the city's workweek was the celebration of Earth Day, an unprecedented national event that was part seminar, part carnival, and part environmental protest. Conceived only months earlier and organized on a shoestring by a recent college graduate, Earth Day was celebrated by 20 million participants nationwide. Congress adjourned to allow members from both parties to speak at college campuses or to march in demonstrations. Government business was put on hold as federal employees participated in teach-ins or activities within their departments. It was the nation's largest environmental demonstration ever.

Like New York's celebration, Earth Day observances in most communities were eclectic. In Philadelphia, 25,000 people attended a teach-in at the University of Pennsylvania where Senator Edmund S. Muskie (Democrat of Maine) called for "an environmental revolution." In West Virginia, volunteers collected five tons of litter and deposited it on the courthouse steps in Clarksburg. In schools, regular classes were replaced with environmental education. The National Education Association estimated that 10 million public school children participated in the day's activities.

For demonstrators across the country, the common enemy was American industry, the source of much of the pollution that people could see or smell. In New York, the city assigned extra police to the Union Square headquarters of Consolidated Edison, viewed as a major polluter. In San Francisco, demonstrators calling themselves environmental vigilantes poured oil in a pool at Standard Oil offices to protest frequent spills. At the Department of the Interior in Washington, D.C., 2,500 demonstrators chanted "stop the muck" to protest oil-drilling leases. In Charleston, South Carolina, a shrimp trawler set off for Washington, carrying petitions with 35,000 signatures to protest the construction of a chemical plant. In Detroit, Michigan, 40 picketers called attention to water pollution from the city's steel mills. Many protests were directed at the auto industry, the target of pollution control rules enacted five years earlier. In New York, Mayor Lindsay asked city employees to give up the use of cars for the day and rode the subway himself. In Tacoma, Washington, 100 students rode on horseback down a freeway to protest auto pollution.

For all its success, Earth Day was an unlikely event. It had been suggested by Senator Gaylord Nelson (Democrat of Wisconsin) and other environmental leaders in Congress only months earlier. Borrowing its teach-in format from demonstrations against the Vietnam War and organized with a budget of $125,000 by Denis Hayes, a 25-year-old Stanford University graduate, Earth Day might well have turned out to be a modest student protest. But the idea tapped into an undercurrent of growing public concern about the environmental damage that had accompanied a generation of extraordinary prosperity. For four months, it was promoted by Hayes, who grew up in Camas, Washington, where his father worked for paper mills, and a group of youthful volunteers. They formed Environmental Action, Inc., and operated from cluttered offices in Washington, D.C., using rented typewriters and copying machines.

Politicians peppered their remarks with references to more prominent issues of the day. Senator Muskie reminded the crowd in Philadelphia that "we are spending 20 times as much on Vietnam as we are to fight water pollution." Senator Birch Bayh (Democrat of Indiana) proposed a federal agency "to conquer pollution as we have conquered space."

There were also critics. Interestingly, Earth Day's most vocal detractors were political leaders who complained that the celebration robbed more pressing priorities of the attention they deserved. Speaking in Washington, D.C., Mayor Richard G. Hatcher of Gary, Indiana, charged that "the nation's concern with environment has done what George Wallace was unable to do: distract the nation from the human problems of the black and brown American, living in just as much misery as ever."

Pundits would argue for decades about whether Earth Day 1970 marked the culmination of years of growing environmental concern or the start of a new era. It was both. In the 1960s, national laws had encouraged states to regulate air and water pollution, had set pollution limits for new cars, had set guidelines for protecting endangered species, and had protected millions of acres of federal land as wilderness. But the slow pace of change, made more frustrating by visible signs of environmental degradation, helped fuel a sudden increase in the public's demand for tougher laws and tighter deadlines. In 1965 the environment was not a leading issue. Five years later, it was the national problem Americans said they worried about most, second only to crime. Earth Day 1970, celebrated just as that crescendo in public concern was reaching its peak, became a lasting symbol both of past frustrations and of future hopes.[1]

The generation of national efforts to protect the environment that followed the celebration of Earth Day in 1970 represents a rare and remarkable achievement in American government: the successful introduction of a new theme into national policy. New themes are unusual in our political system. Sudden change is intentionally minimized by the Constitution's separation of powers and by a legislative structure that encourages the balancing of opposing interests.[2] Yet in the four years beginning in 1969, almost overnight in political terms, congressional action transformed what had been mainly state and local housekeeping chores into a national campaign to protect the environment. Legislation created new public participation in major government and private decisions, new notions of federalism, and new government power over big business.

Those laws also have sparked 30 years of political battles. Advocates of the rapid introduction of new pollution-control technology for factories, utilities, and sewage treatment plants have been pitted against those who have feared that such changes would threaten jobs, competitive advantage, or economic growth. Battles also have divided people who fought for new ecological values in the management of public lands and those who favored more traditional notions of efficient resource use.

As those conflicts continue, the United States has embarked on a much more difficult and politically treacherous environmental campaign— though without the fanfare that accompanied federal efforts in the 1970s. In the last decade, ad hoc responses to novel conflicts have begun to define policies to control many dispersed, often invisible sources of pollution and to conserve ecological resources on private land. Those decisions—and the ones that will follow—are far more complex than choices made during the past 30 years to control large sources of pollution and conserve ecological resources on public lands, difficult as those were. They create new tension between national goals and state preferences, between public concerns and private property, and between common interests and commercial enterprise.

Emerging issues challenge settled habits as much as existing technology, and uses of private land as much as management of national forests and wilderness areas. They raise questions about whether farmers, small business owners, and developers will change their practices, whether drivers will change their commuting habits, and whether homeowners will replace leaking septic tanks, fertilize their lawns differently, or reduce their use of energy. And they raise questions about whether and how landowners will protect ecological resources on patchworks of farms, housing developments, and commercial property while sustaining economic growth. To paraphrase the comic strip *Pogo,* we have seen the enemy and it is us.

4

Scientists and policymakers agree that these are not marginal questions. Today's most serious water pollution threats come from chemicals and organic wastes in farmers' fields, commercial developments, city and suburban streets, and homeowners' lawns, driveways, and septic fields, washed by rain into rivers, lakes, and bays. Today's most serious air pollution threats arise from diverse sources inside homes, schools, and workplaces, in part because that is where people spend most of their time. Today's most serious auto pollution problems are caused by emissions from aging or poorly maintained cars, exacerbated by the fact that drivers log in more miles per vehicle each year. Today's most serious problems from especially hazardous air pollutants can be traced to an array of sources: vehicles and motorized equipment, consumer and commercial solvents, dry cleaners and other neighborhood businesses, as well as factories.[3] Today's most serious conservation challenges arise from individual choices in the use of private property: the ways that farmers work their land, ranchers graze their herds, and developers plan subdivisions. In terms of contentious problems, the direction of change is from the visible, concentrated, and well known toward the invisible, diffuse, and unfamiliar.

Such controversies change the political calculus. They are difficult to resolve through national standards, which have been the usual means of translating national priorities into requirements for business or guidelines for the management of public lands. They sometimes fail to elicit broad public support. In the early 1970s people became outraged by the pollution from factory smokestacks that blackened window sills and reduced visibility, and by the pollution from refineries and steel mills that clogged rivers in some urban areas. It is easier to ignore invisible gases from small businesses, emissions from aging automobiles, and chemicals carried away by storm runoff. They are hard to identify and hard to trace to their source. Also, such issues create new and potentially explosive clashes between the public's proven concern with environmental protection and its commitment to other enduring values: protection of private property, encouragement of small businesses and family farms, preservation of local autonomy in land use decisions, and respect for individual choice.

To complicate matters further, the costs and inconvenience of making such adjustments tend to fall directly on more people than has been true of environmental improvements in the past. American businesses pay a high price for pollution control, more than $76 billion a year.[4] In the abstract, voters may understand that ultimately that price is paid by consumers, shareholders, or employees. Yet the impact of those control efforts is rarely discernible to individual buyers, investors, or workers. By contrast, environmental progress that relies on getting people to maintain their cars or improve their septic tanks is experienced directly by millions of people across the country.

These problems also confound governmental processes. They clash not only with traditional values but also with goals of federal agriculture, transportation, and housing laws, and they confuse lines of federal and state authority. Their complexity and political friction have already contributed to nearly a decade of congressional gridlock, raising concerns about whether the nation has reached a stalemate rather than a turning point in environmental policy. Three laws that help form the foundation of federal mandates—water pollution, endangered species, and hazardous waste disposal statutes—are many years out of date and were due for reauthorization in the early 1990s. Only two major laws, concerning pesticides and the safety of drinking water, have gained congressional approval in recent years.

In one sense, the growing national concern about these more complex, less visible environmental problems is good news. They are receiving more attention in part because 30 years of national efforts have produced clear successes in controlling many concentrated sources of pollution and in improving balanced management of federal lands. These successes are particularly impressive because they have taken place during a time when economic activity has more than doubled, the nation's population has increased by nearly a third, and vehicle miles traveled have increased 121 percent.[5]

The fact that today's most serious water pollution problem is runoff from farms and city streets is a tribute to those successes. Cities no longer dump raw sewage into lakes and rivers. Steel mills, refineries, and chemical companies exercise better control over discharges from their drainpipes. Indoor air pollution is beginning to receive more attention in part because atmospheric concentrations of major air pollutants regulated by the government decreased by a quarter or more from 1977 to 1996, despite economic growth. Reductions include a 97 percent decrease for lead, 61 percent for carbon monoxide, and 58 percent for sulfur dioxide. By the mid-1990s, Southern California, long plagued by the nation's worst smog, had the cleanest air in 40 years, with smog alerts reduced from 121 in 1977 to seven in 1996. Federal regulation today concerns itself with politically volatile issues such as cutting pollution from old cars in part because new cars are at least 90 percent cleaner than those produced in 1970.[6] And issues of conserving ecological resources on private lands have gained prominence in part because public lands, once managed mainly to maximize their yields from logging, mining, and grazing, are now managed to sustain multiple uses, including recreation and habitat protection.

The core idea that pollution control and conservation are appropriate concerns for national action is now accepted by Democrats and Republicans alike. Perhaps the most significant change since the 1970s has been a national learning process that has accompanied years of conflict. Today, Americans are as far removed from congressional action in the early 1970s as policymakers then were from the start of World War II. College students who participated in the first Earth Day demonstrations are 50 years old. Their children, the first generation to be schooled in environmental science and environmental law, are taking their places in the work force.

Environmental politics has matured. While debate in Washington sometimes remains locked in outdated ideological terms, contentious issues around the country are often resolved pragmatically. Neither businesses nor government agencies nor environmental groups speak with a single voice. Alliances break up and re-form from one controversy to another.

Assimilation of environmental values into the political system does not mean an end to aggressive lobbying and big-money politics, of course. Battles continue to be fought over what actions to take, who will benefit and who will pay, how fast to require improvements, and how much of the taxpayers' money to spend. Congressional candidates continue to run against overregulation, overspending, and bureaucratic bungling. But they do not run squarely against the idea of national efforts to improve the environment.

Successes also do not mean that factory pollution is a thing of the past. Manufacturers still discharge millions of pounds of pollutants into the air, on land, or into water—legally. To cite one example, in 1995 industry discharged 2.2 billion pounds of chemicals included in the federal Toxics Release Inventory, most of it into the air.[7] Technological change will continue to play a central role in reducing industrial pollution. But that role, too, is being re-cast. Ongoing efforts to capture pollution from smokestacks and drain pipes before it enters the air or water

7

are beginning to be accompanied by longer term efforts to re-organize industrial processes to minimize waste. Such attempts to employ "industrial ecology" borrow a lesson from nature. Interdependent plants and animals use waste from one process as energy for another. New emphasis is on myriad changes that can reduce use of materials, encourage less-polluting forms of energy, or employ discharges as resources.[8]

Big businesses are still caught cheating. In the largest enforcement action ever under federal air pollution rules, seven manufacturers of diesel engines agreed in 1998 to pay more than $1 billion in fines and corrective actions for allegedly equipping trucks with devices to defeat pollution control equipment. The EPA estimated that, as a result of the settlement, nitrogen oxide emissions from diesel engines would be reduced one-third by 2003. Earlier in 1998, American Honda Motor Company agreed to pay $267 million and Ford Motor Company $7.8 million for selling cars whose engines defeated emission control devices once on the road.[9]

Public lands also remain political battlegrounds. In a confrontation that produced a small, feathered symbol for a decade of conservation efforts, logging interests and environmentalists fought in the 1990s over whether to limit timber harvesting in old growth forests in the Northwest, virtually all of which are on public lands. The U.S. Forest Service had long promoted logging in such forests. The occasion for the controversy was the decline of the northern spotted owl, a "threatened" species under the federal Endangered Species Act, whose habitat includes such forests. After years of negotiation, a compromise reduced logging and offered more than $1 billion in assistance to displaced loggers. Ranching, logging, and mining interests also fought successfully to preserve favorable terms for their activities on public lands.[10] But during a period when the voters' confidence in their public servants has been low, it is worth acknowledging a governmental success—the addition of a new set of values to the American political system, and their influence not just on government and industry practices, but also on reducing pollution and improving conservation.

As less visible and more diffuse problems gain prominence, simple assumptions that formed the foundation for national policy 30 years ago have been replaced with new paradoxes. The framework of the laws enacted between 1969 and 1973 reflected a particular moment in American history. Congress responded to the public's sense of crisis. It responded to voters' enduring trust in its ability to find quick remedies

8

combined with their suspicions that bureaucrats would thwart those efforts. It responded to doubts that state and local governments could or would address pressing problems. And it reflected the public's abiding faith in the capability of big business to find technological solutions to the nation's problems coupled with a deep distrust of its will to use that capability to improve the environment.

A generation later, the political and economic ground has shifted. Basic questions recur and their answers today bear little resemblance to assumptions that have supported national policy for the past 30 years. The public's sense of crisis has been replaced with enduring support for improving pollution control and conservation, but also with a frequent reluctance to pay the public costs of increased protection or to change everyday habits. The federal government's ability to understand and manage environmental problems has increased immeasurably since the early 1970s. But, in practice, possibilities for national action are now constrained by the increasing power of international forces, the declining influence of federal agencies over state and local government actions, an unmanageable workload, and an aging system of laws and regulations sometimes out of sync with new science and new issues. State governments have gained in competence and been given greater responsibilities. But the strength of state programs varies enormously. And states are caught in a funding squeeze between taxpayer revolts and diminishing federal funds. That funding squeeze is especially damaging in less affluent states, where environmental programs may be weaker than in more prosperous states. American companies, once insulated by tariffs and technological superiority, compete in a fast-changing national and international economy. Interestingly, business attitudes toward environmental protection have become more positive as business's economic position has become less secure. But big business also has new reasons to minimize costs, and many smaller businesses and farmers, upon whom much environmental progress depends, lack the resources and incentives to modify their practices.

These changes in public support and in government and business capabilities matter because, under our system of government, approaches to environmental protection, like approaches to other domestic problems, are inevitably collaborative. Federal officials cannot themselves reduce pollution or improve conservation, except on government property. They can make rules, impose penalties, and spend money. But ultimately they rely on the cooperation of state and local governments, businesses, and individuals to produce the results they want. Collaboration becomes more important as public attention turns to environmental problems that are hard to manage with simple national rules.

So far, most national debate about environmental policy has focused on the critical task of making the framework of laws enacted in the 1970s and 1980s work better. Thirty years of legislative initiatives, regulatory actions, and judicial decisions have turned each major law into a web of complex rules and procedures. Finding ways to improve the effectiveness of those laws is of first importance.

Less attention has been devoted to understanding changes in the character of problems drawing new public attention or in the political and economic context for action. This book explores those issues. Its four chapters are an attempt not to foretell the future, but to discern patterns in changes that have already taken place or are under way, and to suggest their possible significance.

Chapter 1 sketches differences between today's environmental problems and those that were the focus of national laws in the 1970s. It explains that increasing public concern about more diffuse and complex problems coincides with mounting evidence that they create serious environmental health and ecological risks.

Chapter 2 examines the foundation for the past 30 years of environmental policy. It describes the chance convergence of political and economic forces that led to the enactment of national laws in the early 1970s. It also attempts to explain how prevailing assumptions of that time shaped national laws with uniform national requirements, precise rules and procedures, demanding deadlines, attempts to minimize administrative discretion, and adversarial proceedings.

Chapter 3 explores how that foundation is now shifting, exposing new paradoxes. After a generation, pollution control and conservation have been assimilated into the American political system. As national priorities, they have stood the test of time, and they have weathered political challenges. They have become a permanent part of government and business decisionmaking. But changing times also have revealed new puzzles about the character of public support and the capacity of government and business to address a new generation of environmental problems.

Policy approaches taking shape against this background are the subject of chapter 4. Because pressing conflicts between environmental protection and economic development must be resolved somehow, federal and local officials, business executives, environmental groups, and voters have engaged in a pragmatic search for means of tackling new problems. Many aspects of national policy are, in practice, becoming customized by state, locality, industry, or facility. The simple structure of uniform standards and deadlines of the 1970s is evolving into a complex web of requirements and negotiated agreements tailored to suit particular situations. New conflicts between federal and state governments are producing some shifts in responsibility. Financial incentives that raise the cost or rewards of private actions have moved into the mainstream of national and state policy. Tax provisions and subsidies designed to raise the priority of environmental protection are being layered on top of provisions designed to foster economic growth, creating confusion. Emissions-trading systems have been adopted in several situations in an attempt to lower the costs of pollution control. What has emerged is a new style of environmental politics.

The search for pragmatic approaches also has highlighted the importance of developing accurate and objectively interpreted information to support decisions and has suggested that information sometimes can serve as a regulatory tool. Fortuitously, exponential growth in computer and communication technology creates unprecedented opportunities to expand that potential. But, so far, national resources have not been deployed to take advantage of information's new power.

This book has a limited purpose. It is neither a history of U.S. environmental policy nor a complete account of federal and state laws and institutions. Many capable writers have taken on those tasks, and their works are readily available for readers who want to pursue those subjects.[11] This work is intended to explore the impact of changing problems and changing times on national environmental policy, in the hope that public and private actions will respond to future needs, rather than to past fears.

1

We Have Seen the Enemy and It Is Us

ON AUGUST 22, 1997, seven doctors from the U.S. Centers for Disease Control and Prevention, Johns Hopkins University, and the University of Maryland traveled to the Eastern Shore town of Westover, Maryland, to examine people who were suffering from skin lesions, breathing problems, and memory loss after wading or fishing in the Pocomoke River, a tributary of the Chesapeake Bay. The suspected cause of their problems, and of massive fish kills earlier in the month, was a mysterious microbe, *Pfiesteria piscicida*, that had suddenly multiplied in Chesapeake tributaries polluted with heavy runoff after an unusually rainy year.

Ironically, the Bay was polluted not with industrial toxins, but with too much of a good thing. Chicken farmers who populate Maryland's Eastern Shore fertilized their fields with chicken manure or commercial fertilizers, both rich in nitrogen and phosphorus. Washed into the Bay during storms, these nutrients set off a destructive chain reaction. Sometimes they fed giant algal blooms. Decaying algae removed oxygen from the water and released toxins, creating dead zones where plants and fish could not live. Occasionally, excess nutrients triggered a population explosion of toxic microbes, among them *Pfiesteria*. In short, one of the chemical heroes of increased agricultural productivity—more fertilizer—had become a villain in the struggle to control water pollution. In all, 24 people were thought to have been harmed by the *Pfiesteria* outbreak, which also killed 30,000 fish. Maryland governor Parris N. Glendening closed parts of three rivers and a creek flowing into the Chesapeake Bay to fishing, boating, and swimming. Uniformed natural resource officers warned people to stay away.

National news stories compared the alarming outbreak in Maryland to other pollution-related fish kills and pointed out that such incidents were increasing in number. In 1987 North Carolina's shellfish industry lost $26 million when shellfish beds were closed for six months because of a sudden takeover by a toxic microbe. Another outbreak in 1995 killed 14 million fish. The same year, more than 25 million gallons of liquid hog manure—twice the amount of oil that the Exxon *Valdez* spilled into Prince William Sound—poured into North Carolina's New River when a holding lagoon at a large farm was breached. In an extraordinary action, the state responded to these incidents by halting the growth of its hog industry, the second largest in the nation. The legislature placed a two-year moratorium on new farms. In 1997, when 100 of the state's residents were thought to have been sickened from a *Pfiesteria* outbreak that had killed millions of fish, *U.S. News and World Report* called the microbe "the cell from hell."

By mid-September two mainstays of Maryland's economy were at war with each other. The portion of the economy that was based on recreation and commercial fishing took a nose dive. Boat sales and rentals declined, and seafood market signs boasted that their fish came from North Carolina. Public officials and businessmen blamed farmers for upsetting the Bay's biological balance by fertilizing fields with chicken manure.

Chicken farmers, another significant economic interest, counterattacked. James A. Perdue, chairman of the state's largest producer, Perdue Farms, called the outbreak coincidental and insisted that farmers could not afford to change their ways. He also pointed out that the state could not afford to lose their business, and that times were already precarious enough for chicken farmers. Maryland's poultry industry, which was producing more than 300 million chickens a year valued at $1 billion, was already hurt by high feed prices and sagging exports.

National, state, and local officials tried to calm public fears while they pieced together a plan of action. The governors of four states surrounding the Bay met in Annapolis with the head of the U.S. Environmental Protection Agency to consider possible actions. Governor Glendening asked former governor Harry Hughes to head a commission to recommend state measures, and a number of public officials engaged in conspicuous consumption of local seafood.

Within two months, federal and state representatives proposed a solution that offered farmers a combination of incentives and requirements to change their ways. They redirected national and state agricultural subsidies to pay farmers to grow buffer zones of grass and trees along 5,000 miles of Bay shore within five years. Such borders, scientists explained, reduced pollution by slowing runoff and drawing off nutrients. The $250 million proposal was part of the ongoing federal Conservation Reserve Program, an effort started in the 1980s to shift some farm subsidies from production toward environmental objectives. It would pay landowners yearly rent for up to 15 years for conserving land bordering the Bay, with the state adding funds to extend some payments. Federal officials declared the plan a milestone in using incentives instead of regulations to meet national water quality goals.

Governor Glendening added more incentives and a strict requirement, devoting a large part of his "State of the State" address to the problem in January 1998. Noting that the mysterious *Pfiesteria* was expected to return in future summers, the governor offered tax credits to help farmers meet the expenses of replacing chicken manure with commercial fertilizer. But he also asked the legislature to approve a new law to place legal limits on the use of chicken manure.

The issue created political rifts not only between the part of the state's economy dependent on fishing and recreation and the part dependent on chicken farming, but also between small farmers and large producers within the poultry industry, and between farmers and suburbanites. The governor's plan concentrated financial burdens on small farmers rather than large producers such as Perdue and Tyson Foods. Owners of small, often low-income farms served as suppliers to these major companies, contracting for supplies and chicks, using manure to enhance harvests of feed corn, and returning grown chickens for a small profit. When the legislature limited the use of chicken manure in May 1998, farmers accused suburbanites of demonizing agriculture and called for controls on fertilizer use by homeowners as well. Legislators said they had avoided that "hot-button issue" to get the measure passed.

Other causes of nutrient buildup received less attention, although the state did commit funds to upgrade sewage treatment plants that discharged waste into the Bay. These plants, along with factories, contributed about a third of nutrient buildup. (Farms accounted for about one-third of nitrogen and more than half of phosphorus; lawns contributed an estimated 10 percent of nutrient buildup.) But there was little discussion of findings by the governor's commission that 6 percent of the Bay's nitrogen load came from faulty septic tanks that leaked 40 million gallons of sewage daily into groundwater or the fact that more than 20 percent of nitrogen originated as air pollution. There was also little attention to the fact that the federal government itself owned 2.2 million acres of land where runoff reached the Bay. The Maryland state legislature approved the proposed plan, and the public awaited the arrival of another summer on the Bay.[1]

Farmers, Small Business Owners, Drivers, and Homeowners

Maryland's ongoing battle against a mysterious microbe is emblematic of the stark contrast between environmental problems that dominated debate 30 years ago and those that produce widespread conflict today. In the 1960s and 1970s visible signs of pollution from factory and government smokestacks and drainpipes and of ecological problems from timber-cutting, mining, and grazing on public lands ignited a crusade to enact powerful national laws. Today chemical and biological interactions that are often invisible and imperfectly understood—the cumulative consequences of thousands of small and productive actions by farmers, developers, small business owners, drivers, and homeowners—kindle more complex political issues.

Much government effort continues to be directed toward controlling large sources of pollution and improving the ecological health of the third of the country's land owned by government. But federal, state, and local officials also contend with water pollution problems caused by urban and agricultural runoff and leaking underground tanks, and with air pollution problems caused by neighborhood businesses, aging automobiles, and substances inhaled in offices and homes. They wrestle with ecological problems caused by developers' plans and farmers' practices. Scientists and policymakers agree that these issues present some of the most serious environmental risks to human health and to ecology.

Why are such controversies gaining prominence now? One reason is past successes. Some visible, concentrated sources of pollution have been reduced. Some guidelines for conservation on public lands have been adopted. Another reason is that new scientific findings continue to extend the known chain of environmental consequences from the immediate and visible hazards toward remote, invisible, and indirect ones. A third reason is that economic and population growth produce more frequent clashes with public demands for environmental protection.

These issues share several characteristics that distinguish them from the kinds of problems that propelled national action in the 1970s. Today's complex controversies often concern public interest in everyday uses of land, cars, or buildings. Frequently, pollution accumulates from many small and dispersed sources located in urban, suburban, and rural areas where most people live. Also, environmental effects tend to be indirect, are often invisible, and are sometimes delayed—as they were in the case of the *Pfiesteria* outbreak in Maryland. In addition, the important actors tend to be ordinary people—farmers, developers, owners of small businesses, drivers, and homeowners—rather than corporate titans or government officials. And though the issues often include elements of national interest, they are also locally unique. Finally, these problems cannot be solved solely by introducing new technology, as pollution problems from industries and new cars often can be. Their solution involves changes in habits, changes in business and farming practices, and changes in private land use. Those changes, in turn, create novel clashes between the public's proven support for environmental protection and its enduring respect for private property, small businesses, family farms, and individual choice.

Water Pollution from Farms and Lawns

Today's most serious water pollution problem in the 40 percent of the nation's waterways that remain unsafe for swimming or fishing is runoff from farmers' fields and city streets.[2] Rain carries the leavings of daily life—fertilizer and pesticides from fields, golf courses, and lawns; oil from driveways and streets; sewage from leaky septic tanks; and the land itself—into rivers, lakes, and bays.[3]

These unremarkable individual events cause serious collective consequences. By changing the biological balance in waterways, such pollution interferes with commercial fishing, recreation, and tourism, as evidenced by recent microbial attacks in Maryland and North Carolina, and by nearly 2,500 beach closings or advisories in 1996 due to runoff.[4] Runoff draining into the Mississippi River from thousands of farms and towns produces a 7,000-square-mile "dead zone" off the coast of Louisiana each summer, where shrimp and fish cannot survive.[5] Agricultural runoff is the leading cause of impaired water quality in nearly half of the nation's lakes and rivers, and urban runoff is the leading cause of water quality problems in estuaries.[6]

Urban and rural runoff also threatens drinking water. *Cryptosporidium*, the microbe that caused the worst documented incident of waterborne disease in U.S. history, washes into lakes and rivers from cattle manure and septic tanks, as well as from sewage treatment plants. *Cryptosporidium* contaminated the water supply in Milwaukee, Wisconsin, in April 1993, sickening an estimated 400,000 people, hospitalizing 4,000, and contributing to the deaths of 50 with weakened immune systems. Lake Michigan is the drinking water source for Milwaukee, Chicago, and many other midwestern cities.[7] As a result, water pollution control has become increasingly a matter of land use control. Leaking fuel storage tanks, often buried underground on farms or at neighborhood gas stations, also threaten groundwater. The EPA has evidence of 360,000 leaking tanks nationwide, with half of the leaks seeping into groundwater. Under rules developed in 1988, farmers and small business owners are required to fix leaks or replace tanks.[8]

Progress in controlling large sources of water pollution has made small sources more prominent. That progress has been harder to assess than the impact of measures to control air pollution because of weaknesses in monitoring information. Even so, there is no doubt about the big story. Cities and towns no longer dump raw sewage into lakes, rivers, or oceans, and they have vastly improved sewage treatment with the assistance of an enormous investment of federal funds, estimated at $66 billion.[9] Discharges from factories into lakes and rivers also have improved, and new plants have to show that they can meet national and state emissions requirements. Surprisingly, water use also declined 9 percent from 1980 to 1995, mainly because of gains in efficiency, despite population growth of 16 percent during the same period.[10]

But there is also cause for continuing concern about industrial and municipal pollution. Freshwater and coastal ecology may have deteriorated during the past 30 years, and sewage spills remain the main reason for closing beaches and restricting shellfish harvesting. Toxic pollution—caused by chemicals that have serious health effects in small amounts—remains pervasive and has many sources, from large factories to small businesses to boats and jet skis. Federal and state laws attempt to limit such pollution but do not ban it. In 1995 industry reported discharging 136 million pounds of toxic pollutants directly into lakes, rivers, and coastal areas. In that year, public officials issued fish consumption advisories for 1,740 bodies of water, with mercury and polychlorinated biphenyls (PCBs) prominent among the causes of concern. In the mid-1990s toxic organic chemicals, mainly PCBs, still contaminated nearly all the Great Lakes shorelines.[11]

Scattered Sources of Air Pollution

Today's most serious air pollution problems include emissions from building materials, equipment, and naturally occurring gases inside homes and workplaces; chemicals from gas stations, dry cleaners, photo processors, and other neighborhood businesses; and exhaust from old or poorly maintained cars, trucks, and buses. According to the White House Council on Environmental Quality, "recent evidence indicates that air in homes, schools, and work places can have higher levels of pollution than outdoor air, affecting health, comfort, and productivity." Pollution in homes and workplaces includes lead, radon, smoke, volatile organic compounds, combustion gases, and particles. Little is known about the health effects of low-level exposure to these pollutants indoors, but the council reports that potential long-term effects can involve asthma, respiratory disease, damage to the brain and nervous system, and cancer.[12] In 1993, when the EPA polled its regional representatives about the relative seriousness of environmental problems, indoor radon and indoor air pollution other than radon were the human health risks most widely rated as "high." The EPA's Science Advisory Board also highlighted their importance as health risks in a 1990 report.[13]

One source of indoor air pollution—secondhand smoke—became a contentious political issue in the 1990s and was aggressively regulated by federal, state, and local governments. By 1996 nearly all states and more than 560 localities had passed laws restricting secondhand smoke in public places, and some states regulated smoking in private workplaces as well. Federal rules restricted smoking in government buildings and on domestic flights and bus trips.[14]

A series of studies of urban and suburban residents in 14 states starting in the 1980s found that the greatest exposure to toxic substances occurs in homes, offices, and cars. Study results suggested that "the exposure arising from the sources normally targeted by environmental laws—Superfund sites, factories, local industry—was negligible in comparison."[15] Secondhand tobacco smoke and radon, a naturally occurring invisible, odorless, radioactive gas, are thought to be the most harmful indoor air pollutants, followed by chemicals from building materials, office equipment, and household products, though scientists still disagree about precise risks. The EPA classifies radon as a known human carcinogen and considers it the second leading cause of lung cancer, after smoking. It is common in soil and rock and seeps into homes through foundation cracks and pipe openings. In 1998 a National Research Council reassessment of radon risks concluded that nearly all of the 19,000 lung cancer deaths each year related to radon (of a total of 160,000) occurred among people who both smoked and suffered exposure to radon. Reducing indoor air pollution means asking property owners to undertake such measures as testing the air, repairing structures, improving ventilation, or changing equipment.[16]

Now that air pollution from large industrial and municipal sources and from new cars has been tamed somewhat, further progress in reducing atmospheric pollution depends heavily on changes made by farmers, and by owners of small businesses and of aging cars, trucks, and buses. According to government estimates, the combined emissions of the six major air pollutants under federal regulation decreased 32 percent between 1970 and 1996, despite economic and population growth, a notable accomplishment. Concentrations of carbon monoxide, lead, and sulfur dioxide are less than half as great as they were in 1977. National and state controls, the decline of heavy industry, and changes in weather patterns contributed to the decline.

Most metropolitan areas that still do not meet federal air pollution standards are experiencing persistent problems with controlling ozone and particles. In 1996, according to EPA estimates, 46 million people still lived in areas where at least one pollutant exceeded national standards. Both ozone and particulate standards were further tightened in 1997 in response to the EPA's updated assessments of health and environmental risks posed by those pollutants. Ozone is associated with decreased lung function, increased susceptibility to respiratory disease (especially in children), and reduced crop and forest yields. Coarse particles, measuring between 10 and 2.5 micrometers in diameter and long regulated by the EPA, are believed to aggravate respiratory problems. Fine particles, measuring 2.5 micrometers or less and newly regulated, are associated with more serious respiratory problems, though the science remains controversial.[17]

These two pollutants have proved to be particularly difficult to control. They are generated by thousands of diverse sources and millions of cars, as well as by industry. Ground-level ozone, still pervasive and the main component of smog, results from a complex interaction of sunlight with nitrogen oxides and volatile organic compounds. According to the EPA, power plants are responsible for about a quarter of nitrogen oxide emissions. Transportation sources account for nearly half. About 42 percent of volatile organic compounds come from cars and trucks, and about 50 percent from industrial processes, mainly those that use solvents.[18]

Particulate pollution, long regulated by EPA, results almost entirely from the ways people use their land and from natural causes. Direct emissions are chiefly associated with agriculture, forestry, managed burning, and fugitive dust from roads, as well as natural processes such as wind erosion and wildfires. According to the EPA, the leading sources of finer-particle pollution, recently regulated, are motor vehicles, power plants, industry, and home fireplaces and wood stoves.[19]

Scientific advances have also caused concern about the toxic effects of some chemicals used in everyday life. Those that cause hazardous air pollution (that is, pollution that is known or suspected to cause severe health effects) are distinguished in federal law from the six common air pollutants that are nationally regulated. Congress specified 189 hazardous air pollutants from diverse sources for federal control in 1990 amendments to the Clean Air Act, including heavy metals, organic chemicals, and pesticides. Sources include hundreds of thousands of small businesses. Dry cleaners, users of solvents, commercial sterilizers, and chrome platers, for example, are prominent among the EPA's top 20 sources of toxic emissions, which account for 79 percent of total toxic air emissions. On-road motor vehicles and residential wood-burning are rated as the largest sources of toxic pollution.[20]

Research results in the 1990s raised new toxicity questions about whether some of the 87,000 chemicals in everyday use might also be endocrine disrupters—compounds that interfere with human hormones and thereby affect reproduction, growth, and brain chemistry. In 1996, prompted by such studies, Congress called for a screening process for common chemicals. After two years of study, a committee of experts recommended that the EPA begin an unprecedented program of analyzing the possible hormonal effects of 15,000 of the most widely used compounds, a program the EPA initiated in August 1998.[21] Responding to these signs of concern, chemical companies announced in January 1999 that they would invest $1 billion in a six-year research project to determine the safety of thousands of common substances.[22]

Motor vehicle emissions still account for about 40 percent of the pollution that contributes to ozone in the lower atmosphere, 41 percent of toxic air pollution, and 60 percent of carbon monoxide emissions (95 percent in cities). But now that new cars are at least 70 percent cleaner than they were in 1970, maintenance and driving decisions by owners of older cars, trucks, and buses account for most such pollution. (New trucks, buses, and sport utility vehicles remain major polluters, however, because they are subject to less strict emission standards than are cars. In the northeastern United States, for example, heavy-duty trucks produce a third of nitrogen oxide emissions and often have faulty pollution control devices, according to the EPA.[23])

The increasing prominence of controversies concerning pollutants from small and diverse sources does not mean that problems from industrial or government pollution have been solved. Air and water pollution laws grant industries permits that are, in effect, licenses to pollute, albeit within set limits. In 1995 industry emissions of toxic chemicals included in the federal Toxics Release Inventory amounted to 2.2 billion pounds, with more than two-thirds emitted into the air. Industry contributes much of the 750 million tons of hazardous waste are generated each year, and air pollution emissions that in 1996 included 89 million tons of carbon monoxide, 23 million tons of nitrogen oxides, and 19 million tons of volatile organic compounds and of sulfur dioxide.[24]

Ecological Issues on a Patchwork of Private Lands

The steady march of urban and suburban development and specific farming techniques cause some of today's most troubling ecological conflicts. Some issues stem from a clear conflict between economic and environmental interests: farmers' interests in maximizing acres under cultivation versus environmentalists' interests in maintaining buffer zones along lakes or rivers, for example. Others illustrate that those interests can be intertwined in complex ways. Businesses and communities near the Chesapeake Bay and the Florida Everglades, for instance, rely heavily on those aquatic resources for their economic health, to provide jobs and revenue for commercial fishing, tourism, and other businesses. In many localities across the country, trees and other vegetation play a critical role as natural pollution filters, protecting the quality of water for drinking. Concern about global warming adds a new layer of complexity to ecological issues. Scientists believe that preserving and expanding forests may be a cost-effective means of reducing the amount of carbon dioxide in the air, and thus of reducing the risk of climate change. Rotating corn crops with soybeans while cutting down on chemical fertilizers may also reduce carbon dioxide significantly.[25]

As science has advanced, national policy has moved toward protecting broad areas of ecological importance rather than individual endangered species. The Department of the Interior now negotiates habitat conservation plans with private landowners and engages in complex land swaps to preserve large swaths of land considered ecologically important. Government agencies and private environmental groups are in the early stages of trying to develop a science-based consensus identifying ecosystems that are particularly threatened or that should be protected from development (see chapters 3 and 4).[26]

But good science makes for contentious politics. Protecting larger land areas means trying to influence plans of more landowners and more state and local jurisdictions. That shift in emphasis raises novel issues about the future development of urban, suburban, and rural land. In the area around San Diego, California, for example, efforts originally directed at protecting the habitat of a specific endangered species—the California gnatcatcher—grew into a general five-year debate about development versus the ecological protection of hundreds of thousands of acres in urban, suburban, and rural areas. In Portland, Oregon, efforts to protect the habitat of a threatened fish called the steelhead grew into a broad controversy about protecting the entire lower Columbia River Basin, involving questions about lawn care and home building, as well as sewage treatment and industrial discharges.[27]

Decisions about development and the use of natural resources may be influenced by government policies, but in the end they are made mainly in private markets. About 72 percent of the land in the contiguous 48 states is privately owned, and the proportion is much higher in many areas of rapid growth.[28] An estimated 70 percent of critical habitats and half of the nation's wetlands are located on private land.[29] In 1990 the EPA's Science Advisory Board, attempting to assess the relative importance of the nation's ecological risks, found habitat alteration and wetlands loss to be two of the most serious risks.[30]

Continued population and economic growth preordains problems. From 1970 to 1996, the U.S. population increased by almost a third and economic activity more than doubled. Although the population is expected to grow at a slower rate for the next 60 years, it is still projected to reach 347 million by 2030. Between 1980 and 1992 land in urban areas increased by a quarter, to nearly 60 million acres. Acreage used for crops, range, and pasture is in steady decline, though production has grown, thanks to improvements in plant and animal breeding and to the use of fertilizers and pesticides.[31]

Plants, animals, and humans often compete for the same small portions of a vast country, intensifying environmental pressures. Interestingly, it is not the rising proportion of land used for housing and commercial developments in the nation's total land area that poses problems. Developed land still amounts to only about 6 percent of total nonfederal land. Rather, it is competition between development and conservation interests for limited acreage highly valued for both purposes. Coastal areas, for example, are particularly important ecologically. Coastal wetlands are essential to the life cycle of many fish and birds, and coastal vegetation filters polluted waters from cities and farms. But coastal areas, including those bordering the Great Lakes, appeal to home buyers and to many businesses as well. In 1994, 53 percent of the nation's population lived in coastal areas that accounted for 11 percent of the land area. More than half of the population increase between 1960 and the mid-1990s was in those areas, as was nearly half of all building construction between 1970 and 1989. Population density along the Pacific Ocean and the Gulf of Mexico nearly doubled between 1960 and 1994.[32]

As the country becomes more crowded, government officials attempting to resolve such issues are turning to the daunting task of persuading ordinary people to change their habits, and big developers and agribusinesses to alter their planning. That often means modifying long-established practices associated with some of the country's proudest achievements, including improving agricultural productivity, fostering home ownership, and increasing mobility. As discussed in chapter 3, this is politically treacherous terrain.

In traversing this territory, politicians are attempting to patch together new accommodations between their constituents' fervent and incompatible wishes. In particular, voters would like to improve attention to environmental consequences of land use while continuing their long romance with the American farmer. For 200 years the image of the family farmer tilling his fields has been a touchstone of American independence and productivity. It remains one of the nation's cherished myths.[33]

Times have changed, however. Today, 54 percent of farming land consists of large farms (with annual sales of $100,000 or more and an average of 1,542 acres). These operations account for 83 percent of farm produce and 98 percent of poultry sales, and they are twice as likely to receive government subsidies as are smaller farms. Farmers' use of chemicals has doubled in the past 30 years. In 1995 American farmers applied 566 million pounds of chemicals to their lands, including 324 million pounds of herbicides and more than 200 million pounds of pesticides.[34]

Yet farmers remain exempted from most environmental rules that apply to other businesses. Under the Clean Water Act, for example, nearly all of the nation's 1.1 million farms are excluded from water pollution rules as "nonpoint" sources. Only large animal feeding operations—perhaps 10,000 in number—are considered point sources covered by the law.[35]

Old Problems, New Prominence

Most of these widespread pollution and ecological problems that have new prominence on the public agenda are not, in fact, new. Many were debated in the 1970s and 1980s, and some were included in national laws, often with a call for improved understanding or planning. But all have been neglected in practice in favor of more visible, more easily manageable, and more politically palatable problems.

National water pollution laws and agriculture laws from the 1970s on called for public and private planning to address the problems of urban and rural runoff and included modest grant funds, but such pollution was not targeted in most national regulations, and state regulation has been variable.[36]

Air pollution rules have called for programs of inspection and maintenance for cars on the road since 1970, but driver resistance has slowed action in many states.[37] Indoor air pollution has been little understood and has received relatively little attention from national and state governments. J. Clarence Davies of Resources for the Future has pointed out that "[i]ndoor radon is not regulated by EPA at all and only minimally by the states. The EPA radon program is 0.36 percent of the agency's budget. . . . Similarly, indoor air pollution . . . is for the most part neither regulated by EPA nor by the states. . . . EPA has an indoor air program—it is 0.15 percent of the agency's budget."[38] National efforts to improve the ecological health of private lands in the 1970s and 1980s were generally limited to voluntary programs aimed at farmers or requirements for specific geographical areas such as coastal counties, wetlands, and habitats of endangered species. Even so, they often produced explosive confrontations (see chapter 3).

Not Business as Usual

These newly prominent issues confound the normal workings of government.[39] Many of them are not amenable to resolution using the usual regulatory techniques. Dispersed sources of pollution and opportunities for conservation may not lend themselves to uniform standards, the most common means of dealing with federal priorities.

Also, the growing national interest in the environmental consequences of land use threatens the power of local zoning, planning, and health authorities. In the United States, land use decisions have traditionally been a local matter. Local agencies are the traditional arbiters of conflicts between public needs and landowners' plans on questions about water supply, sewage treatment, or preservation of open space, or about the health and safety of building materials, or the need to test for toxins or to provide for ventilation.

Furthermore, these newly prominent issues challenge the long-standing legal and bureaucratic barriers that separate pollution control programs from ecological programs and that treat air, water, and land pollution as distinct concerns. Over the years, federal and state governments have addressed environmental problems one at a time, constructing individual regulatory frameworks to deal with air, water, and land pollution. The cumulative result has been a proliferation of environmental baronies, located in various agencies, departments, and congressional and state legislature committees.[40] In the federal executive branch, for example, most pollution issues are assigned to the Environmental Protection Agency, while programs concerned with ecological health are divided among the Department of the Interior (for public lands and endangered species), the Army Corps of Engineers (for wetlands), and the Department of Agriculture (for farms). But emphasis on urban and rural runoff and other new issues blur those distinctions. By encouraging farmers to grow grass and trees to filter runoff, for example, the state of Maryland uses ecological measures as a means of pollution control. Likewise, programs that are designed to protect wetlands and watersheds, or that rely on natural biological processes to reduce air pollution or to reduce pollution that moves among air, water, and land, straddle established regulatory lines.

Environmental remedies for newly prominent problems also collide with long-standing goals of other federal and state programs. Since the Great Depression, farm programs have tended to discourage production in order to stabilize prices. In response to meager world harvests in the 1980s, however, agricultural programs gave farmers incentives to plant marginal lands, including wetlands and acreage bordering waterways.[41] After World War II, the GI Bill of Rights and federal mortgage support encouraged the construction of new homes. Automobile sales boomed, and federal highway funding created the interstate highway system and made it possible for states to build new roads in undeveloped areas. These practices combined to underwrite unprecedented residential and commercial development. Between 1945 and 1955, 15 million housing units were constructed.[42]

The conflicts now surrounding these issues make it more urgent to understand how assumptions about public support for environmental programs and about government and business capabilities have changed. The next chapter takes up that question.

2

Yesterday's Agenda and Where It Came From

ON MARCH 21, 1969, President Richard M. Nixon took a 90-minute aerial tour of a massive blowout from a Union Oil Company well located off the coast of Santa Barbara, California, a disaster that newspaper headlines called "the accident that experts said wouldn't happen." At its worst, the blowout covered more than 400 square miles of water with a 6-inch layer of crude oil, blanketed at least 30 miles of beaches with 200,000 gallons of stinking ooze, killed thousands of sea birds, wiped out nearly all local fishing for weeks, and brought a local economy 80 percent dependent on the tourist trade to a sudden halt. The spreading puddle of oil, from a well in the ocean floor that leaked for more than a week before company employees brought it under control, provided television viewers across the country with repeated reminders that government and industry had failed to prevent a disaster. Protesters on the gooey beach stuck their oil and gas credit cards on barbecue skewers and toasted them. Union Oil president Fred Hartley was booed by picketers when he arrived to tour the damage. He later told a Senate subcommittee that he was "always tremendously impressed at the publicity that [the] death of birds receives versus the loss of people."

Leaders of both political parties deplored the incident and competed to promise forceful action. Senator Edmund Muskie (Democrat of Maine), candidate for vice-president in 1968 and chairman of the Senate's Public Works subcommittee concerned with pollution, was the leading environmental advocate in a Democrat-controlled Congress. He toured the area shortly after the accident, promptly held hearings, and promised to introduce legislation to hold companies responsible for leaks and spills. California's Republican governor, Ronald Reagan, demanded that drilling be stopped and asked for stronger federal regulations; Santa Barbara's political leaders joined the appeal. Nixon, the newly inaugurated Republican president following eight years of Democratic rule, promised city officials that he would consider their request to stop oil drilling in the channel and, a few months later, declared the 1970s "the decade of the environment." As the White House Council on Environmental Quality reported 27 years later, the Santa Barbara blowout showed that the "federal government had largely ignored the need to protect commercial, recreational, aesthetic, and ecological values of the area."[1]

The nation's waterways seemed awash in spilled oil in 1969, reflecting in part the media's growing interest in industrial pollution and adding to the public's concern that government and industry were not protecting people from environmental hazards. In January an oil spill suspected to be from a tanker coated 20 miles of beaches on Long Island Sound.[2] In May another spill threatened those beaches again.[3] In June a leaking barge spread oil for 15 miles on the Mississippi River.[4] In September a barge hit a rock and spewed oil into Buzzard's Bay off Cape Cod, Massachusetts.[5] But the incident that produced the most dramatic image of the year was Cleveland's burning river. The Cuyahoga River bubbled with pollution and was said to be incapable of supporting any form of life. In June the surface of the river and a trestle of a railroad bridge caught fire, apparently when sparks from a passing train ignited oil-soaked debris floating beneath the bridge. The bizarre event gained national attention, and the river "burned on newscasts all over the world," according to a Cleveland State University professor. It became "a vivid symbol of the state of many of America's waterways."[6]

Four Remarkable Years

The images of 1969 featured spreading oil slicks and factory wastes dumped into rivers, and public lands stripped by timber-cutting or mining, delivered to American living rooms via newly popular color television. Together, they helped fuel growing public outrage and demands for national action. That groundswell combined with competition between a Democratic Congress and a Republican White House and with the new political savvy of environmental organizations to propel to passage an extraordinary group of environmental laws between 1969 and 1973. Bipartisan majorities in Congress approved major statutes to combat large, visible sources of pollution and to give new priority to ecological concerns, primarily on public lands. These laws included measures to control air and water pollution, create national parks and wilderness areas, protect endangered species and marine mammals, protect wild and scenic rivers and coastal areas, regulate pesticides, and integrate environmental concerns into major federal actions.

Congressional action was based on four ideas that reflected the political and economic forces of those years and that have influenced environmental policy ever since:

—public perception that there was an environmental crisis requiring immediate action;

—confidence that Congress could solve environmental problems, coupled with growing suspicion that bureaucrats might thwart those efforts, not because they lacked skill but because they could be captured by interests they were supposed to regulate;

—distrust of both the capabilities and the political will of state and local governments to reduce pollution and improve conservation; and

—faith in the apparently unlimited ability of big business to meet new challenges and especially to improve technology, combined with skepticism about executives' will to reduce pollution.

Upon this foundation Congress constructed a complex framework of environmental laws. The laws responded to the public's sense of crisis by demanding "major action throughout the Nation, . . . major investments in new technology, . . . [and restriction of] as much as seventy-five percent of the traffic" in some metropolitan areas. Leaps in technology were required because "the health of people is more important than the question of whether . . . [change] is technically feasible," as the Senate committee report on 1970 air pollution legislation put it.[7] Laws often limited agency discretion by specifying amounts and timetables for reducing emissions of pollutants. State and local flexibility was limited not only by uniform national rules but also by deadlines and criteria for implementation plans, and by strengthened federal enforcement. Congress also authorized citizen suits to help "motivate governmental agencies."[8]

Those four ideas need to be revisited after a generation of environmental progress (see chapter 3). But, first, how did they become prominent and why have they had such lasting power?

A Chance Convergence of Political Forces

Competition between two politicians propelled the issue of environmental protection to the top of the national political agenda. Ironically, each had reasons to *resist* a move toward federal requirements. Edmund Muskie had spent much of his career fighting to preserve state prerogatives. Richard M. Nixon, elected president by a narrow margin in 1968, promoted a sweeping plan to return power to state and local governments. Nonetheless, their two-year competition in preparation for the 1972 presidential election provided a powerful political force in favor of strong national laws.

Senator Muskie, who had turned an undesirable assignment to the Public Works Committee into a command post for environmental action during the 1960s, believed that the federal government should support, not seize, the power of states to fight pollution. He knew the complexities of economic development and environmental protection issues firsthand, having grown up in a small paper mill town in the traditionally business-dominated state of Maine and having served as the state's governor. He had helped to modify congressional proposals in the 1960s, arguing that pollution control should remain primarily a state and regional matter, should be weighed against other economic and political needs, and should be phased in gradually. Through most of the 1960s, he did not favor national standards for industry.[9]

Then, on the eve of the 1970 midterm elections, the 56-year-old Muskie consolidated his position as Nixon's most likely challenger for the presidency in 1972. In a 15-minute, nationally televised address watched by nearly 30 million people, he responded in measured tones to Nixon's harangue against student demonstrators protesting the Vietnam War. Muskie asked the American people to replace the politics of fear with the politics of trust.[10]

But Muskie himself was attacked by newly powerful environmental groups for his moderate views. Ralph Nader's two-year-old Center for

Study of Responsive Law, one of the leading groups, sponsored an attack on him in *Vanishing Air*, a widely publicized book charging that Muskie had "failed the nation in the field of air pollution control legislation."[11] Sparring with Nixon for national leadership and responding to the public's growing sense of crisis, Muskie began to promote pollution laws with strict national standards, tight deadlines, and little provision for weighing other national priorities.[12]

Republican Richard Nixon built a political career fighting big government, but in 1970 he also needed to seize the political initiative from Muskie. A Democratic Congress had buried his program to make government work better by creating superdepartments to consolidate the management of major programs. Democrats rejected his innovative family assistance plan to replace welfare with a guaranteed minimum income for all families. And his moves to end the Vietnam War were obscured by secrecy, mishaps, and popular protest. His nationally televised attack on the eve of the 1970 congressional elections, berating the "thugs and hoodlums" who protested against the war, backfired. Democrats increased their majority in the House by 9 seats, retained control of the Senate, and added 11 governorships in off-year elections. Muskie, the Democrat chosen to respond to Nixon, gained in the polls.[13]

Elected with only 43 percent of the popular vote in 1968, Nixon needed to take bold steps to expand his ideological base in order to be reelected in 1972, and he needed to counter Muskie's appeal. After the 1970 elections, the president had to find something "elevating" to talk about, his speechwriter Ray Price explained.[14] Democrats controlled both the House (243 to 192) and the Senate (57 to 43). Nixon would have to play on their terms. The idea of saving the environment also had captivated many Republicans, some of whom had roots in the conservation movement that ran back to Theodore Roosevelt. And action in several states to set diverse pollution standards, led by California, brought calls from some businesses with national markets for federal regulation, as a defensive measure.[15]

Nixon fought Muskie for environmental leadership for two years while Muskie gained in the polls. The president supported the National Environmental Policy Act requiring that federal officials consider environmental consequences of major federal actions, created a presidential Council on Environmental Quality, and established the Environmental Protection Agency. He devoted most of his 1970 State of the Union address to an ambitious program to fight pollution and promote conservation, which included enactment of national air pollution standards and water pollution criteria and the creation of a federal Department of Natural Resources. By the time Muskie dropped out of the presidential race early in 1972, the nation had acquired major elements of a new framework for promoting environmental protection.[16]

National Policies Have Deep Roots

The national environmental policies that seemed to spring suddenly into being in the early 1970s were in fact deeply rooted in a long history of national concerns both with public health and with conservation, a fact that helps explain why they have stood the test of time and of political challenge. These two concerns became intertwined in the 1950s and 1960s as worries about the side effects of industrial growth gained political ground. Until the 1970s, though, pollution control was primarily a state and local task, supported by national research and by some national funding, and conservation efforts were directed mainly toward improving the productivity of public lands.[17]

Public Health Returns to Environmental Concerns

A key development in the 1960s was the emergence of national concern about environmental health. Interestingly, that issue brought the focus on public health full circle. Initial interest in environmental causes of ill health in the nineteenth century, gradually replaced by attention to specific disease-causing pathogens as bacteriology advanced in the twentieth century, now widened again to include pollution concerns in the 1960s. Paul Starr writes in *The Social Transformation of American Medicine* that as cities grew in population in the nineteenth century, public health was mainly a matter of sanitation. When scientists discovered that disease was caused by germs, not by general impurities, attention shifted from the environment to the individual. However, public health remained primarily a matter of state and local action until after World War II, when the National Institutes of Health and the Public Health Service came to play a leading role in medical research.[18]

By the 1960s national concern about public health returned to a question raised more than 100 years earlier: what harm did the ordinary surroundings of everyday life impose on human health? With the ben-

32

efit of improving science, public attention turned to cigarettes and sweeteners, and to air and water pollution. Writing in 1966 about changes in the postwar years in *The Great Leap,* John Brooks described a view held by René Dubos, among others, that freedom from disease was illusory because "victories over bacterial disease . . . were concurrent with the rise of potentially deadly air pollution and indiscriminate use of poisonous insecticides."[19]

Interest in Air Pollution Control Rekindled

While aggressive regulation of chemicals in air and water was not on most state and local agendas in the 1960s, pollution control had periodically been a local issue for growing metropolises. Many cities adopted smoke ordinances in the nineteenth century. As Chicago prepared to host the World's Fair in 1893, smoke from factories and coal-burning furnaces blackened clothing and stung people's eyes. Newspapers declared that "the smoke nuisance must go," and business leaders formed a Society for the Prevention of Smoke to inspect buildings and persuade owners to switch from soft coal to cleaner-burning anthracite and to install pollution-control devices.[20]

Air pollution control gained new public support in booming industrial cities after World War II. Led by Los Angeles, many cities enacted local ordinances to control soot and smog, sometimes building on the earlier laws that declared dense smoke or other pollution a public nuisance. In 1948 a "killer smog" in Donora, Pennsylvania, linked to 20 deaths and 6,000 illnesses, drew national attention to the consequences of pollution for human health. Corporate tsar Richard Mellon and Mayor David Lawrence spearheaded Pittsburgh's successful effort to reduce air pollution beginning in 1946 as part of their plan to revive the city's economy. And in the 1950s, the federal government devoted substantial funds to research on the health effects of pollution.[21]

National action began to play a central role during the 1960s as public health concerns grew. In 1963 Congress passed the first major national air pollution law, the Clean Air Act, encouraging regional cooperation, providing technical assistance, and giving state and local governments grants for pollution-control initiatives. Congress authorized national auto pollution standards in 1965 and required states to set air pollution controls and submit plans to carry them out in 1967. Widely publicized incidents increased the public's understanding of the links between pollution and health. In November 1966, for example, a stagnant mass of polluted air hung over New York City for four days and was blamed for 80 deaths.[22]

Water Pollution Control Targets Commercial Needs and Epidemics

Early measures to control water pollution were driven by commercial needs and by public health concerns. The Rivers and Harbors Act of 1899, which required a federal permit for the discharge of refuse into navigable waters, was intended to remove hazards to commercial ships making their way through crowded channels.[23] Epidemics of waterborne diseases and disputes among upstream and downstream cities about sewage dumped in rivers made water pollution a pressing concern for growing cities at the end of the nineteenth century and led to local ordinances. Postwar migration to urban areas strained drinking water and sewage treatment systems and led to state action. By 1948 all states had water pollution control agencies. But funding the construction of sewage treatment plants proved difficult, especially because cleaner water often benefited mainly the residents of downstream communities.[24]

In the 1950s Congress provided increased funding for treatment plants on the condition that states adopt pollution-control plans. Congress also backed state and local efforts with federal research and technical assistance and authorized national action when interstate water pollution threatened public health. By 1968 roughly half of state agency funds and half of treatment plant funds were coming from the federal government.[25]

By the mid-1960s, however, Congress had become impatient with the slow progress of the states. The Water Quality Act of 1965 gave federal authorities the power to set water quality standards for interstate waterways if state standards were inadequate. At the same time, Congress increased funding for local sewage treatment to record levels, authorizing $3.5 billion for construction in 1966 legislation.[26]

Conservation of Resources Focuses on Public Lands

National interest in conservation on lands owned by the federal government began with a focus on productivity in the early 1900s. This grew to include a concern with preserving wildlife habitats fostered by hunters, fishermen, and bird-watchers in the 1920s and 1930s and expanded to take into account pollution control and broader ecological interests in the 1960s.

After a 100-year effort by the federal government to give away public lands to anyone who would take them, the conservation movement of the early 1900s began striving instead to manage them more efficiently. Championed by Republican president Theodore Roosevelt, the movement was initially a reaction against the wasteful use of water, forests, and other public resources on the third of national acreage that had fallen into federal hands a century earlier, mainly because of agreement among eastern states to give up western claims and the Louisiana Purchase.

Conservationists, whose spiritual leader was Gifford Pinchot, a pioneer in professional forestry, favored building dams and draining wetlands to conserve freshwater and managing national forests to produce sustainable yields. They vied for political power with preservationists, led by John Muir, founder of the Sierra Club, who successfully lobbied for the creation of national parks. By the 1930s, preservationists' interests in protecting wildlife habitats had been reinforced by sportsmen's organizations such as the Izaak Walton League and the National Wildlife Federation.[27]

The activism of the 1960s gave rise to political battles over the "New Conservation," as Stuart Udall, secretary of the interior in both the Kennedy and Johnson administrations, called it, adding more interests to be protected on public lands. In a time of growing prosperity, millions of American families bought cars and had the leisure to explore the nation's natural wonders. Their numbers made Washington more receptive to protecting mountains and forests for recreation. The Multiple Use, Sustained Yield Act of 1960 required federal agencies to manage national forests to serve many interests, including "recreation, range, timber, watershed, and wildlife and fish purposes."[28] A 1962 law named recreation as a secondary purpose of wildlife refuges. The Wilderness Act of 1964 set out specific standards and procedures for managing designated land, and a 1968 law raised the importance of recreation uses of protected rivers.

Nonetheless, both recreation and the protection of wildlife habitats continued to be considered subordinate to the primary goals of the Department of the Interior and the Department of Agriculture. During the 1960s and 1970s managers of public lands and public works projects continued to see economic growth as their mission, despite a growing list of competing interests. Congress acknowledged those interests by telling agencies that had previously had a relatively free hand in deciding what to do to consult with other agencies and to consider recreation and wildlife concerns. In planning water projects, the Army Corps of Engineers had to take into account the views of the Fish and Wildlife Service. Other agencies planning development projects were required to respect "social and cultural values." Federal highway officials were instructed to make special efforts to prevent harm to parks, recreation areas, and wildlife refuges.[29]

As the 1960s drew to a close, the groundwork for the next stages of pollution control and conservation had been laid. National concern about environmental problems was growing, as was frustration with the slow progress in addressing them. But a chance combination of political interests and economic circumstances made these next stages—a series of landmark laws passed by Congress between 1970 and 1973—more than a predictable step forward. They were a surprising leap.

The Public Demands Action

The idea that postwar prosperity and population growth had increased environmental hazards was not new in 1970, nor was all pollution getting worse. Such data as there are indicate that emissions of sulfur oxides, nitrogen oxides, volatile organic compounds, and carbon monoxide had increased between 1960 and 1970. At the same time, particulate emissions had actually declined, mainly as a result of municipal bans on garbage and leaf burning and the phasing out of coal as a fuel for heating.[30]

What was new was the public's sense of urgency. Accidents like the Santa Barbara oil spill contributed to the idea that government and big business had failed the American people. Improving science increased understanding of the links to human health and of the damage to nature. Skillful advocacy by newly prominent environmental groups, some of them steeped in the spirit of 1960s activism, increased public support and put pressure on Congress to act. Also new was the emerging idea that something could be done about it all, and done quickly. Despite the growing skepticism of the time, voters still believed that national laws could solve problems, if they were framed so as to avoid bureaucratic maneuvering and lobbying by industry.

In 1965 few people considered pollution to be an important public issue. Five years later it was a leading concern, second only to crime.[31] Membership in environmental groups jumped from about 124,000 for the 12 largest groups in 1960 to more than 800,000 by 1969 and more than a million by 1972. In 1973 half the environmental groups surveyed by the federal Council on Environmental Quality had been formed in the preceding four years.[32] The leaders of these groups tended to be well educated and relatively affluent, but a broad spectrum of the public supported environmental protection, according to various polls.[33] By April 22, 1970, when ordinary Americans, including 10 million school children, gathered all over the country to celebrate the first Earth Day, a majority of those polled supported a strong national program to reduce environmental damage.[34] As Theodore H. White wrote in *The Making of the President 1972*, "The environment cause had swollen into the favorite sacred issue of all politicians, all TV networks, all good-willed people of any party."[35]

These shifts in public opinion did not take place in isolation from other events. An increasingly prosperous middle class demanded not only the cars, refrigerators, and washing machines that were driving economic growth and increasing pollution. As voters, they also demanded political reform. They expected professional managers to replace political bosses and smoke-filled rooms, and they expected to participate in decisions. They used taxpayer suits to challenge proposed highways, dams, and city-renewal efforts. Public works projects that had been routine in the past became venues for battles among dozens of separately defined and legally empowered political interests.[36]

Democrats and Republicans alike competed for the votes of growing numbers of suburbanites. Prosperity created mobility. More people rejected conditions in the central cities, including pollution, and moved to rapidly growing suburbs. The Supreme Court's 1962 decision in *Baker v. Carr* began a process of redrawing the boundaries of congressional and state legislature districts to reflect postwar population shifts to urban and suburban areas, giving suburbanites greater political power.[37]

Improving pollution control and conservation was not a partisan issue. President Nixon, whose administration proposed strong laws and reorganized the executive branch to give environmental concerns greater prominence, was a Republican and so were 32 governors. Tough new laws were approved without significant opposition. The House passed the 1970 Clean Air Act by a vote of 374 to 1. The Senate approved a somewhat different version of the law unanimously. Even conservative politicians championed the cause. Governor Reagan of California called for "an all-out war against the debauching of the environment" and proposed the most ambitious air and water pollution control standards in the country.[38]

Environmental science, though still in its infancy, contributed to the new sense of urgency in two ways. First, it showed that seemingly local pollution could damage human health and plant and animal life far removed in time and place. As Rachel Carson explained in *Silent Spring*, the most widely read environmental book of the time, the DDT used to control the spruce budworm turned up in fish 30 miles away and affected reproduction in the birds that ate the fish.[39] Exhaust from cars, trucks, and buses was linked to urban smog, which in turn contributed to respiratory problems. Mercury was found in fish in many lakes and rivers, far from its suspected sources.[40]

Second, some scientists argued that seemingly separate problems could have cumulative effects that might threaten the future of the nation, and perhaps the world. Not only did auto emissions combine with chemicals from power plants and coal-burning home furnaces to blanket cities in soot and smog. Widely read scientists such as Paul Ehrlich predicted (wrongly, as it turned out) that world population growth was outstripping the resources to support it. Those concerns were reenforced by the Club of Rome's *The Limits to Growth* report in 1972, predicting that growing population and industrialization could lead to environmental disaster by the end of the century.[41]

Much remained to be understood and measured. Automobiles were identified as one of the causes of air pollution only in the 1950s. Even states that had air pollution laws in the 1960s did not have reliable means to measure it, and about half the states did no monitoring at all. For scientists, understanding pollution and conservation was, and continues to be, a tough assignment. Causal connections are complex, controlled tests are difficult, and consequences are sometimes many miles or decades removed from their triggering events.[42]

Two unrelated and seemingly obscure rule changes, one by the federal courts and one by the Internal Revenue Service (IRS), increased the power of environmental groups to alter national politics. First, beginning in the mid-1960s federal courts gradually changed their rules about who could bring lawsuits, allowing members of the public, and groups representing them, "standing" to sue companies without showing economic or physical injury. Traditionally, courts heard only those cases in which there was lost revenue, harm to health, or other specific damage to meet the Constitution's "case or controversy" requirement. This procedural change, specifically authorized in most environmental statutes and further broadened in the early 1970s, allowed environmental groups to represent public interests in challenging companies' practices or government agencies' adherence to congressional mandates.[43]

Second, the Internal Revenue Service decided in 1969 to allow environmental groups to keep the tax-exempt status of foundations and charitable organizations while they sued companies and lobbied Congress. That meant they did not need to pay income taxes, could receive tax-exempt contributions from foundations and individuals, and could benefit from low postal rates. The change took place in two steps: the IRS, under pressure from environmental advocates and sympathetic members of Congress, ruled that "public interest law firms" could keep tax-exempt status; then, under the Tax Reform Act of 1969, the IRS allowed environmental groups to do limited lobbying and to set up related organizations to influence legislation. These actions provided the economic underpinning for the environmental movement. They also created the opportunity for environmental groups to use direct-mail solicitation, a cheap way to reach many people and to attract members.[44]

These opportunities coincided with evolutionary changes in environmental groups themselves, and in the way they were funded. In the 1960s, traditional conservation advocates began to reflect the postwar migration to metropolitan areas. They added urban issues and concerns about nuclear power to their agendas and hired professional staffs. These groups were joined by new and politically savvy organizations whose goals, tactics, and targets differed from those of the Izaak Walton League or the Audubon Society. New groups such as the Environmental Defense Fund and the Natural Resources Defense Council, each with their own focus, linked human health concerns to pollution, used lawsuits and dramatic actions to make their case, and targeted big business. Their efforts were underwritten by large grants from the Ford Foundation and other national givers.[45]

The public's understanding of environmental problems influenced the character of the political response. Action focused on large, visible sources of pollution, including natural resource industries, factories,

and local government sewage and waste disposal facilities, and on opportunities for conservation on public lands. Air, water, and land pollution appeared as separate problems in incidents that drew public attention, and they were separately addressed in new laws. Furthermore,
the public expected big business—which had, as people saw it, created
the problems—to pay the bill for the cleanup.

Hope and Skepticism Shape Federal Remedies

By the late 1960s confidence in the power of national government to
solve health and safety problems had become tempered by skepticism
about whether agencies would follow through on congressional promises. It was also tainted by the cynicism spawned by the Vietnam War.
In 1964 and 1965 many people believed that the government could help
rebuild cities, reduce racism and poverty, and ensure minimum health
care for people in need, and that the country was wealthy enough to
pay for these initiatives. The civil rights movement and the "war on
poverty" embodied the idea that the federal government had the ability and the obligation to intervene to protect the rights of the individual,
even if those rights were ignored by any given state. Optimism was
reenforced by Democratic control of Congress, with the result that many
of these convictions were quickly translated into powerful national laws.
Democrat Lyndon B. Johnson, who had assumed the presidency after
John F. Kennedy's assassination in 1963, beat Republican Barry
Goldwater in 1964 in a landslide. The early years of Johnson's presidency, an extraordinary moment of national optimism, produced the
Civil Rights Act, the war on poverty, medicare and medicaid legisla-
tion, immigration reform, assistance for public schools, the Voting Rights
Act, and many other reforms.[46]

Confidence in congressional power was enhanced by the idea that
domestic policy was unaffected by what happened in the rest of the
world. Though the nation was entangled in an unpopular war in a small
country 10,000 miles away, there was no doubt that the United States
could control its own economic and social destiny. Bounded by vast
oceans to east and west and by lesser powers to north and south, the
United States seemed to have complete control over its own domestic
policies. The new cars, refrigerators, and stereos purchased by a prosperous middle class were nearly all made in the United States, the war's

economic effects were not yet fully felt, and other links between the domestic economy and international events seemed minimal.

By the time pollution control and conservation laws were framed six years later, the political climate had changed drastically. Disappointment at the slow progress in carrying out national initiatives and congressional frustration about its limited influence on the U.S. military's actions in Vietnam contributed to growing suspicions that the executive branch could thwart congressional goals, and that bureaucrats could be unduly influenced by special interests. Chronic congressional fears that agency officials would subvert legislators' purposes were heightened by the election of a president from the opposing party. President Nixon himself distrusted career civil servants and tried to minimize their influence. At the same time, television coverage of the Vietnam War provided daily reminders of technology's destructive power. It showed massive environmental damage as well as the loss of human life.[47]

A reaction to the big-spending policies of the New Frontier and the Great Society was also gaining ground. In congressional elections, Republicans picked up 47 seats in the House and 3 seats in the Senate. When President Johnson asked Congress to approve a new list of consumer, health, safety, and city-renewal programs, legislators turned him down. After 1968 the economy also began to show strains of rising wartime inflation, though unemployment and taxes remained low. In *Grand Expectations,* the Oxford history of the United States in the post–World War II years, James T. Patterson, professor of history at Brown University, notes that from the mid-1960s on, smaller numbers of Americans had faith in elected officials or the government, and fewer people turned out to vote.[48]

This distrust of bureaucracy was fortified by the work of a group of political scientists who held that giving bureaucrats broad discretion contributed to the business "capture" of regulatory agencies. In 1969 *The End of Liberalism,* a book by Theodore Lowi that drew on the work of legal scholars Kenneth Culp Davis and Henry J. Friendly, promoted the idea that national regulations should have clear goals and should be accompanied by specific steps to carry them out. Davis believed that "eighty or ninety percent of the impact of the administrative process comes from informal action which is not reviewed." Urging that "we must try to find ways to minimize discretionary injustice," Davis argued that "in such an effort lies . . . the greatest promise for improving the quality of justice to individual parties in our entire legal and governmental system."[49]

States Engage in "Race to the Bottom"?

At the same time, the track record of state and local governments in the 1960s convinced many members of Congress that they lacked the capability and political will to improve pollution control. The national framework of environmental protection constructed between 1970 and 1973 altered the character of federalism by making what had been mainly state and local responsibilities a matter of national requirements. Two ideas justified the change. One was that some pollution had significant interstate consequences, either because it drifted, flowed, or seeped from state to state or because it was associated with nationally sold products, such as automobiles. The other was that states were engaged in a "race to the bottom" to minimize environmental protection in order to attract business.

Events during the previous decade also seemed to bear out the idea that state and local governments were resistant to federal efforts to bring social change. The images of Alabama governor George Wallace standing in the schoolhouse door to block integration of the University of Alabama and of Sheriff Jim Clark's assault on civil rights demonstrators on the Pettus Bridge in Selma were still fresh in people's minds. In the context of congressional initiatives to rebuild cities, improve civil rights, and reduce poverty, many state governments seemed to be remnants of an earlier time when rural interests dominated politics and political patronage determined program managers.[50]

In some states and cities, pollution control was still low on the political agenda in the mid-1960s, though the same forces that produced national action to protect the environment were also at work, often more slowly, in the states. In 1966 federal officials complained to Congress that "not more than half of the urban areas which are in need of . . . programs for air pollution control . . . have them and of these . . . the majority are operated at an inadequate level."[51] In a study of Gary and East Chicago, Indiana, published in 1971, Matthew A. Crenson found that pollution control was kept off the local agenda by politically enforced neglect.

Local groups that stood to gain from growth promoted economic development issues with sustained effort. But no group had enough at stake in pollution issues to give them constant attention. The mayor of Gary did not prepare for an air pollution survey because he was busy with a downtown parking issue. Environmental issues were raised by political outsiders or city officials, not by party regulars. Support was diffuse, and environmental activists were rare. Hence "municipal inaction [was] a regular response to the air pollution problem in communities throughout the nation." For local governments, facing environmental problems also was costly. Their waste-burning incinerators and sewage treatment plants made them major contributors to air and water pollution.[52]

Where pollution was on the political agenda, laws often provided for a cooperative approach to decisionmaking, emphasizing growth and minimizing public participation.[53] Responsibility for pollution control was frequently assigned to part-time boards dominated by industry representatives and economic development interests. In case studies of environmental protection efforts by nine states, funded by the Ford Foundation and published in 1973, Elizabeth H. Haskell and Victoria S. Price found that "in practice, these bodies have been run as the private clubs of the regulated interests."[54] Members were appointed for terms, and interest group representation was sometimes specified by statute. With much at stake, business interests outresearched, outspent, and outlobbied poorly funded and loosely organized groups representing public concerns, whose activism was sometimes constrained by a fear of losing their tax-exempt status. Laws were often administered by state health departments, staffed by sanitary engineers who were charged with carrying out many other laws.[55] Rural legislators, who dominated many state legislatures and were inclined to be parsimonious, tended to oppose spending on air pollution and water pollution problems in distant cities.[56]

Even cities with considerable environmental programs had trouble regulating large industries. Chicago's first comprehensive air pollution law, enacted in 1958, exempted the city's steel mills. A stronger law in 1963 included them, but city air pollution administrators granted the four largest mills eight-year variances from its provisions.[57]

In their successful campaign for national standards, the president, members of Congress, and environmental advocates drew together these strands and concluded that states were inevitably engaged in "a race to

43

the bottom" to minimize environmental standards in order to attract business. In a special message to Congress on the environment in February 1970, President Nixon argued that "without effective standards, states and communities that require such [environmental] controls find themselves at a . . . disadvantage in attracting industry, against more permissive rivals."[58] The House Committee Report on the 1970 Clean Air Act also stated the assumption explicitly: "The promulgation of Federal emission standards for new sources . . . will preclude efforts on the part of States to compete with each other in trying to attract new plants and facilities without assuring adequate control of large scale emissions therefrom."[59]

The phrase harkened back to attempts by state governments to increase jobs during the depression. Mississippi and other states in the South launched "smokestack chasing" campaigns to attract factories from industrial states in the North, using general obligation bonds and tax breaks to help finance the construction of factories.[60] The idea that there was "a race to the bottom" among states implied not just that business was a powerful interest group in state and local politics, a fact that was indisputable, but that an irresistible economic imperative had predetermined the outcome of environmental issues.

It is worth noting, though, that the 1960s were no less a time of upheaval and growth in the states and cities than they were in the nation as a whole. In a long and contentious process set in motion by the Supreme Court's 1962 decision in *Baker* v. *Carr,* state legislatures were redrawing the lines of state and local electoral districts to reflect population shifts. The composition of new districts gave city and suburban voters a stronger voice in statewide issues and elections. Charged by Congress with new responsibilities and armed with new federal resources to rebuild cities, fight crime, and improve education, state and local governments were becoming more sophisticated managers.

In response to fast-changing public opinion and to a decade of pressure from Congress, states had new incentives to better regulate pollution. In 1963, 16 states had air pollution laws; by 1968, 46 had them. Some states moved aggressively to regulate auto emissions. California enacted auto pollution standards in 1960. By 1965 an array of proposed regulations in New York, Pennsylvania, and other state legislatures caused auto manufacturers to endorse the idea of uniform national standards instead of risking diverse state actions.[61]

How far and how fast these efforts would have proceeded without further federal action is a moot question 30 years later. Congressional prodding certainly contributed to state action in the 1950s and 1960s. State governments responded in various ways, and state interests in pollution control had apparent limits (see chapter 3).

What is clear is that the character of national action was influenced by a distrust of state and local capabilities. Congress established national criteria and national deadlines. As a practical matter, laws had to be administered and enforced by state governments. But those responsibilities were circumscribed by state-by-state and program-by-program delegations of authority conditioned on compliance with a variety of specific procedures, by federal monitoring of individual state permit decisions even when responsibilities were delegated, and by continuing federal enforcement.

Big Business Seen as Powerful and Recalcitrant

By the late 1960s the spectacular record of American business growth and technological improvement juxtaposed with the growing numbers of pollution incidents convinced many people that big business had unlimited capabilities but little interest in cleanup and conservation. The incongruity between the power and wealth of big business and its apparent disregard for the health and welfare of ordinary people provided much of the energy to drive national action and contributed to a dual vision. Because of its technological capabilities, big business could be expected to reduce pollution quickly and, for consumers and taxpayers, painlessly, but, because of its intransigence, it would have to be forced to do so by clear, strict national measures backed by harsh penalties.

Within the memory of most voters in 1970, American industry had achieved the impossible, repeatedly. Responding to a sudden and unexpected crisis, it had provided the technology and equipment to win World War II. It had responded to President Kennedy's challenge in 1961 to land a man on the moon by the end of the decade by producing a nationally televised success in 1969. People who had grown up doing laundry by hand and waiting for the iceman's deliveries to keep milk and meat from spoiling delighted in appliances that the large corporations produced, and in the doubling of family incomes from 1950 to

1970 that made such purchases possible. By providing health insurance and pensions, big business had also given American workers a measure of security previously unimagined.[62]

In the late 1960s the wealth and power of business also seemed secure. American corporations orchestrated the march toward greater prosperity, apparently unaffected by events elsewhere in the world. The U.S. economy stood supreme, unchallenged by other countries. International investment, slowed by the depression and World War II, increased enormously in the 1950s and 1960s. But it remained a small part of the total economy and was concentrated among a few big companies. About 200 large companies made more than half of their direct investments abroad. Most American firms remained insulated from foreign competition. High tariffs and superior technology protected many products.[63]

Set against this background of corporate success, incidents such as Santa Barbara's oil blowout and Cleveland's burning river tapped into the populist streak in American politics, with its distrust of big business. The 1960s were a time of particular concern about the concentration of corporate power in relatively few big businesses, its use to influence politics, and its possible destructive force. Economic success seemed to create expanding webs of corporate power among businesses that had no functional relationship, and among tiers of smaller businesses that were suppliers or customers. The growth of the "military-industrial complex" seemed to imply a symbiotic relationship providing mutual gain between big government and big business during the cold war. And the application of that power to a long, unpopular war in Vietnam added to the growing distrust. Corporate secrecy also fed suspicion. Businesses often resisted government attempts to gather information about their activities, including data on health, safety, and pollution issues.[64]

Negative responses by some business executives to early government attempts to improve environmental protection and product safety contributed to the sense that business could not be trusted and sometimes used underhanded tactics. Chemical companies attacked Rachel Carson's credibility after the publication of *Silent Spring* in 1962, helping the environmental cause. The startling revelation that General Motors had hired private investigators to dig up personal information about consumer advocate Ralph Nader after the 1965 publication of *Unsafe at*

Any Speed, his attack on the safety of American cars, suggested industry willingness to misuse its power. And congressional debate about air pollution standards in 1970 took place against the backdrop of a federal suit against auto manufacturers for conspiring to block the adoption of pollution control devices.[65]

Americans were right that the power of big business had increased markedly since the end of World War II. Economic power in the United States was becoming concentrated in the hands of relatively few companies. In his history of American business, *The Visible Hand*, Alfred D. Chandler Jr. explained that by 1968 the 200 largest industrial corporations in the United States held more than 60 percent of manufacturing assets, compared with 47 percent in 1947. He concluded that "as the large enterprises grew and dominated major sectors of the economy, they altered the basic structure of these sectors and of the economy as a whole." The search for new markets in the 1960s led some big businesses to acquire unrelated companies and to form conglomerates.[66]

Laws Reflect the Times

Congress and the president responded to the public's demands. Between January 1, 1970, when President Nixon signed the National Environmental Policy Act, and the end of 1973, Congress approved an extraordinary set of eight major environmental laws, as well as a reorganization plan creating the Environmental Protection Agency. Though new provisions have been added to those laws since then and new priorities have been addressed in separate statutes, the initiatives approved in those three years established a national framework of environmental protection and continue to set many of the ground rules for trade-offs between economic development and environmental protection 30 years later.

Major laws bore the imprint of the time and carried it forward as agencies slowly translated congressional goals into volumes of rules and procedures, and as courts interpreted those rules. The public's sense of urgency produced laws that demanded quick results, targeted mainly large corporations, called for a high standard of protection generally without regard to cost, and emphasized public participation. Response to the visible results of pollution in widely publicized incidents (and

47

the demands of congressional committee structure) helped create a legal structure that treated conservation and pollution as separate problems and viewed air, water, and land pollution as distinct regulatory issues. Belief in the power of national directives coupled with a suspicion of bureaucracy framed laws that demanded bold action but attempted to limit agencies' choices both by legislating specific requirements and, in some instances, by forbidding the balancing of environmental concerns and competing goals. Distrust of state and local governments led to laws that emphasized uniform standards, minimized local differences, and accompanied delegation of responsibilities to states with detailed requirements and oversight. At the same time, public confidence in business power and inventiveness, tempered by suspicion of its will to change, produced laws requiring fast changes in technology and imposing strict rules and harsh penalties.

Three of the most influential laws of those years, which set national rules for controlling air and water pollution and for protecting endangered species, illustrate the ways in which basic assumptions were translated into legal provisions. The Clean Air Act of 1970 called for sharp reductions in major pollutants almost immediately. For example, auto manufacturers were to cut hydrocarbon and carbon monoxide emissions by 90 percent by 1975 (and nitrogen oxides by 1976), the assumption being that new technology could be quickly introduced. The law focused requirements on big business, calling for emission standards for major new sources of pollution to be uniform by industry for all parts of the country. It forbade consideration of economic or technological feasibility in decisions about setting standards (though it allowed such consideration in decisions about how to carry them out). Ambient air quality standards had to be set at a high level to allow an adequate margin of safety to protect public health, including the health of people with respiratory problems. The law allowed federal officials to delegate to states responsibility to carry out its provisions, but hemmed in that permission with a list of specific elements to be included in state plans, continuing federal authority over the review and rewriting of those plans, federal oversight of state decisions, and direct federal enforcement. It also provided for public participation in rule making and in decisions to allow new sources of pollution, and allowed citizens to sue companies or government agencies for violations, and to sue the EPA for failing to carry out the law.[67]

The Water Pollution Control Act of 1972 set a remarkable goal: navigable waters were to be "fishable and swimmable" by 1983 and discharges were to be eliminated altogether by 1985. The act targeted pollution from factory drainpipes and municipal sewage treatment plants and required nationally uniform, technology-based criteria and permits, to be issued source by source. It gave relatively little attention to pollution from farm and urban runoff and other scattered sources. Deadlines were tight. The executive branch was directed to issue criteria for various industries within one year, and to issue permits to each manufacturer within two years. The law required state plans to meet a variety of criteria and relied on a vigilant public, authorizing public participation in rule making and permit decisions, and in citizen suits.[68]

The Endangered Species Act of 1973 also included nationally uniform requirements and provisions to empower the public to counteract bureaucratic maneuvering. It directed federal agencies to ensure that their actions did not jeopardize the continued existence of endangered or threatened plant or animal species, or damage their critical habitats. The law also made it illegal to "take" an endangered animal species. That provision was interpreted by the courts to include harming, killing, and eventually even disturbing the habitat of an endangered species, and to extend to actions on private land. Citizens were authorized to sue any person violating the law and to sue the government for failing to enforce the law.[69]

Other laws took similar approaches. A 1972 law set up the strict regulation of pesticides, expanding registration requirements to include environmental factors and allowing the government to ban harmful products.[70] Another 1972 law banned ocean dumping of waste except by permit in specific locations.[71]

While prevailing assumptions about government and business capabilities dominated congressional action, some laws reflected countervailing views. The National Environmental Policy Act, for example, took a holistic approach. It required the federal government to examine its own major actions for environmental consequences, broadly defined. The Coastal Zone Management Act provided a structure for broad land use planning in coastal counties. Interestingly, a national land use law, promoted by the Nixon White House and by Senator Henry M. Jackson (Democrat of Washington), passed the Senate in both 1972 and 1973 but failed to pass the House in 1974 because of Chamber of

Commerce lobbying and the Nixon administration's loss of interest. The bill would have provided federal grants for state land use plans and for state regulation of critical areas.[72] Both water pollution legislation and a law intended to encourage better practices by cities in disposing of trash authorized extensive federal grants for construction.[73] And, as later experience would show, air and water pollution laws in fact left room for substantial agency and state discretion.

The laws reached far beyond sewer pipes and smokestacks, ultimately changing the way many public decisions were made. They influenced the management of national parks and other public lands. They changed the balance sheets for ranchers, loggers, and mining companies who had built their businesses on prevailing assumptions about the use of public lands. They influenced the design of American cars and the materials and production processes used by chemical companies, steel mills, utilities, and other major businesses. At each step, environmental groups and politically powerful corporations spent millions of dollars fighting in the courts, lobbying agency officials, trying to gain votes in Congress, and making their case to the public. Battles were acrimonious and compromises imperfect. Because most public attention has been drawn to these fights, it is easy to forget about the common ground. Environmental protection has become part of the fabric of American society, its premises supported by Democrats and Republicans alike.

The four ideas that drove congressional actions—the public's sense of crisis, faith in congressional remedies coupled with suspicion of bureaucracy, distrust of state and local governments, and ambivalence about the power of big business—set the tone for an entire generation of environmental policy. Those assumptions still have power in political battles that polarize national debate. But, 30 years later, political and economic forces have in fact transformed the reality. As the nation prepares to tackle new kinds of environmental problems in a new century, there is an urgent need to update understanding of the strength and limits of public support, national initiative, state and local capabilities, and business roles in environmental protection. These revised assumptions are an essential underpinning for the next generation of policy.

3

Today's Puzzling Paradoxes

IDEAS ABOUT PUBLIC SUPPORT and about government and business capabilities that have dominated environmental policy since the 1970s are out of sync with today's anachronisms. The public supports stricter air pollution standards but balks at reducing the driving and energy consumption that contribute to the problem. A once-bold Congress fails to update basic environmental laws due for revision since the early 1990s. Governors of some conservative states take aggressive action to control pollution by one of their major industries—large animal-feeding operations—while governors of others move to contain development. And big business often seems to be at war with itself. Eastern power companies fight for more stringent controls of air pollutants while midwestern companies oppose them. In these puzzling times, basic assumptions that formed the foundation of environmental policy are no longer useful. Federal and state policymakers now grapple with surprising paradoxes:

—The public's sense of crisis has been supplanted by steady pressure for more environmental protection, but also by resistance to changing daily routines or to paying for the monitoring and enforcement required to support improved protection.

51

—The federal government has grown in capability but may be shrinking in influence. Its authority is countered by powerful forces: the increasing role of international politics and multinational businesses, the strengthened voices of state and local governments in national decisions, and an insidious threat of hollow government as commitments far outstrip the resources available to carry them out.[1]

—States are engaged not in a "race to the bottom," but in a fiercely competitive "race to the bottom line" to achieve prosperity. Their initiative in pursuing particular environmental policies varies widely. State governments have vastly greater responsibilities than they did 30 years ago and have found some environmental measures politically beneficial to furthering their aims. But they, too, are burdened with practical limitations. Resources are squeezed between taxpayer revolts and federal budget constraints. And there is growing evidence of a gap between rich states with strong environmental programs and poor states with weak programs, at least at the extremes.

—Oddly, big business's acceptance of environmental requirements has grown during a period when American companies have lost economic security and gained new reasons to minimize costs. But few companies go beyond compliance with government rules or integrate environmental issues into managers' decisions. And pollution and conservation by small businesses and farmers—who have fewer resources and less technical knowledge to make improvements—is a more prominent issue.

Before exploring these paradoxes, it is worth pausing to note that the unique convergence of political and economic forces between 1969 and 1973 that produced a powerful and lasting environmental agenda ended abruptly when hard times prevailed. That sudden change slowed the pace at which the agenda was translated into action and shaped the political battles that accompanied it.

Times Changed Quickly

Even as Congress acted in the early 1970s, surprising events were beginning to occur, limiting congressional initiatives and complicating the task of carrying out national directives already approved. Congress's decisive actions in fact took place in an economy that was deteriorating because of spending on the Vietnam War that had sown the seeds of inflation, budget deficits, and unemployment.

Signs of trouble appeared early. By 1971 fears about growing inflation had led the Nixon administration to take the extraordinary step of instituting a 90-day wage and price freeze. Those fears, reinforced by business protests against the high costs of environmental regulations, also led the White House to add an extra review step for pollution control rules, sending them to other agencies for comment. The White House also ordered a review of the inflationary impact of all major new rules. In his biography of President Nixon, Stephen Ambrose reports that in August 1971 Nixon wrote on a memorandum prepared by his advisers on the economic impact of actions to control pollution, "We have gone overboard on the environment—and are going to reap the whirlwind for our excesses—get me a plan for cooling off the excesses." Ambrose notes that Nixon wanted "credit for boldness and innovation, without the cost."[2]

Political forces also shifted rapidly, punctuated by a chance combination of missteps, violence, and scandal. By early 1972 Nixon no longer needed to compete with Muskie, who had made a weak showing in early primaries. Nixon's main challenger was George Wallace, former governor of Alabama, and then George McGovern, after Wallace was shot while campaigning in Laurel, Maryland. When voters returned Nixon to office, he had less need to compromise. He had won a clear mandate with more than 60 percent of the popular vote and carried every state except Massachusetts and the District of Columbia. Democrats did retain their control of Congress but lost twelve seats in the House.[3]

But the pivotal event of the early 1970s was the sudden shortage and price increases of domestic oil after the Organization of Petroleum Exporting Countries (OPEC) cut off shipments to the United States in October 1973 to protest U.S. backing of Israel in the Arab-Israeli War. The seemingly invulnerable U.S. economy was in fact vulnerable at critical points. It depended on imports, mostly from the Middle East, for one-third of its oil. OPEC's unexpected action triggered shortages, caused lines at gas stations, led some colleges to cancel winter sessions, and prompted Congress to reduce the speed limit on interstate highways to 55 miles an hour. After the embargo was lifted in March 1974, OPEC nations continued to limit production and raise prices. Gas prices in the United States remained as much as 30 percent higher after the embargo than they had been before it was imposed.[4]

For environmental initiatives, the effect was immediate. Congress extended deadlines for reducing air pollution for hydrocarbons, carbon monoxide, and nitrogen oxides, and efforts to reform the American automobile emphasized energy efficiency more than pollution control. (Pollution-control devices had decreased the energy efficiency of cars by about 10 percent between 1970 and 1973.)[5]

By chance, the Arab oil embargo coincided with a domestic political shock that left the White House preoccupied with problems that were more immediate than environmental protection. The day after Saudi Arabia announced a halt of all shipments of oil to the United States, President Nixon fired Archibald Cox, a special prosecutor chosen by the Department of Justice to investigate the president's alleged illegal activities, including his attempts to cover up White House involvement in a break-in of Democratic National Committee offices in Washington's Watergate complex. As the Watergate scandal unfolded, it eroded the president's ability to manage conflicting priorities, including energy conservation and environmental protection, in a time of economic uncertainty.

Even so, congressional efforts to improve pollution control and conservation did not stop. An important national initiative to establish drinking water standards was approved the next year.[6] Several laws gave greater weight to environmental considerations in the management of public lands.[7] New environmental crises, particularly the threat of toxic substances symbolized by reports of chemicals leaking into homes built over a former chemical dump at Love Canal, New York, led to further legislative action.[8]

But after 1973 the public's interest in improving environmental protection was countered by worries about rising prices, growing unemployment, and declining productivity. By 1975 unemployment was running at 9 percent and the nation was in the midst of the worst recession since the 1930s. The "good news years," as Alice Rivlin has called them, were over.[9] Concern about the economy became paramount and added force to industry attacks on the high cost of regulation.

These sudden changes had the effect, over time, of placing in bold relief Congress's actions to control pollution and support conservation measures during the first three years of the decade. Political battles are still fought in the vernacular of the early 1970s, emphasizing national crisis, quick solutions, state recalcitrance, and business bad faith. At the same time, public support, federal initiative, state and local capabilities, and business roles have changed dramatically.

The Enigma of Public Support Emerges

Americans who agree about little else join in their support for environmental protection. In 1995 Everett Carll Ladd, president of the Roper Center for Public Opinion Research, and Karlyn H. Bowman of the American Enterprise Institute analyzed a wide variety of polling data to determine how public attitudes about environmental protection have changed in the 25 years since the first Earth Day demonstrations in 1970. They found that demand for government action to avert a national crisis had been replaced by an enduring commitment to environmental protection based on appreciation of successes to date, optimism that continuing improvements could be balanced with economic growth, and an abiding faith in technological progress. Large majorities said that government regulation was about right or did not go far enough, even though nearly two-thirds rated the quality of air, water, land, and wildlife excellent or good, and most people did not view environmental protection as an urgent issue like crime or drugs.[10]

Yet evidence indicates people are unwilling to change their daily habits in ways that are individually costly or inconvenient, even when the government tells them such changes are required to improve environmental protection. In their review of polling data, Ladd and Bowman found that majorities would not alter their everyday routines and believed that technology would provide environmental progress. Three-quarters of those polled, for example, reported that their households did not change car use for environmental purposes.[11]

After 30 years of political battles, pollution control and conservation have won a permanent place on the American political agenda, a remarkable achievement in a system that is intentionally loaded against major change. Clashes between economic development and environmental protection remain frequent. Conflicts over *how* and *how much* to improve protection and *who pays* are debated in thousands of national, state, and local forums, with powerful interests arrayed against one another. But the debate rarely focuses on *whether* pollution control and ecological health are appropriate subjects for national action. That issue is settled.

Voters may be reluctant to increase government funding for environmental protection, but the nation's commitment is now built into the economy, most of it invisible to voters because it is paid by business. As of 1994 the annual bill for pollution control was estimated at more than $121.8 billion a year, or nearly 2 percent of the nation's gross domestic product, and was increasing at twice the rate of economic growth.[12]

Despite bitter and often paralyzing battles in Congress, the few environmental laws that have managed to run the gauntlet of legislative approval in the 1990s have increased the breadth and depth of national programs. Nearly all Democrats and Republicans voted for the Clean Air Act of 1990, the biggest expansion of air pollution programs since 1970, and one that includes 90 percent cuts in auto emissions.[13] Large majorities also supported new laws that expanded federal requirements to ensure safe drinking water and to control pesticides. Administratively (and under court order), the Environmental Protection Agency significantly tightened controls on fine particles and smog.

Polls indicate that public attitudes toward environmental protection have matured. Ladd and Bowman report that despite declining confidence in government in the mid-1990s, 57 percent of the people thought that the federal government was carrying out its environmental responsibilities fully or fairly well. Large majorities thought industry had made progress in controlling pollution. Most people did not think that economic growth and environmental protection were complementary, but a growing majority shared the conviction that the two could be balanced effectively. In 1992 the Roman Catholic Church declared abuse of the environment a sin.[14]

That said, environmental issues remain extremely contentious. They inevitably pit powerful and well-funded interests against one another. When the EPA proposed major tightening of air pollution rules for fine particles and ozone in 1996, for example, corporate interests and trade associations set up a lobbying organization with contributions of as much as $100,000 from each of more than 500 companies, and support from 15 governors. Opponents charged that new rules costing business $8.5 billion a year would not improve public health and might restrict the use of cars, lawn mowers, fireplaces, and outdoor grills.[15]

Environmental issues also continue to tap into fundamental tensions in national politics between federal and state authority, and between public goals and private choices. Twice, once in the 1980s and once in the 1990s, Republican leaders drew on those tensions to recast environmental protection as a partisan issue. In 1981 Ronald Reagan, a newly elected Republican president who pledged in his first inaugural address to "restore the balance between levels of government," cut spending for environmental programs and attempted to reassign responsibilities from the federal government to the states. In 1995 House Speaker Newt Gingrich led a group of Republicans in the 104th Congress—the first Republican-controlled House and Senate in 40 years—to promise a Contract with America with some of the same themes. Republicans then waged a campaign to weaken federal rules, reduce spending, and rewrite environmental laws.[16]

In both challenges, moves to cut spending and temper federal policy with provisions for greater flexibility served as reminders of the deep divisions in the electorate. But neither changed the direction of national laws, and both ultimately were rebuffed by members of their own party. By 1996 mainstream Republicans counseled fellow members of Congress to reaffirm support for environmental protection. John McCain, Republican senator from Arizona and chairman of the Commerce Committee, warned: "Polls indicate that the environment is the voters' number-one concern about continued Republican leadership of Congress" because Republicans are considered "too eager to swing the meat ax of repeal when the scalpel of reform is what's needed."[17] In the end, those confrontations affirmed that the American people have adopted the broad values of environmental protection, and that those values have been assimilated into the political system, where they compete with other priorities in thousands of federal, state, and local decisions.[18]

The public's growing attention to environmental threats also has made neighborhood facilities that pollute controversial and has raised questions about whether such facilities are disproportionately sited in low-income or minority neighborhoods. In the 1990s advocates gave voice to diverse concerns about environmental risks in minority and low-income communities with a call for environmental justice. Scores of petitions charged that specific state approvals discriminated against minorities.

In 1994 the Clinton administration responded by issuing Executive Order 12898, requiring federal agencies to address "disproportionately high" environmental risks of their policies in minority and low-income communities. As that broad command slowly permeated departmental rules, complex questions arose about its precise meaning. Was business compliance with federal and state rules sufficient to avoid charges of creating disproportionate risks? Were states with formal delegations to carry out federal laws obligated to assess each permit request for compliance with the order's charge? Also, by the late 1990s some analysts were suggesting that calls for environmental justice could be usefully broadened to address the most pressing health risks affecting minority and low-income communities rather than focusing mainly on problems concerning chemicals.[19]

The variability of public involvement may well be the Achilles' heel in a system of environmental protection that relies on citizen participation to assure fairness. Community groups have demonstrated that, when persistent, they can have great influence. Sometimes, they have used environmental laws to drive businesses away. After many dramatic protests, for example, including the blocking of a section of Interstate Highway 95 for several hours, a community coalition in the South Bronx succeeded in July 1997 in getting the city government to shut down a medical incinerator in their neighborhood, citing evidence that pollution from the plant caused asthma and other ailments. Crosscurrents of public opinion that encouraged new uses of abandoned inner-city industrial sites, encouraged concentration of industry and housing to avoid sprawl, and, at the same time, opposed placement of polluting facilities near residential neighborhoods demonstrate the growing complexities of environmental policymaking.[20]

However, experience of 30 years has raised doubts about public willingness to support new rules that require changes in the use of private property or in everyday habits. To date, most progress in pollution control has been the result of technological rather than behavioral change. Most progress in protecting ecological values has resulted from improvements in public rather than private land management.

The rare situations in which federal or state environmental rules have imposed new limitations on landowners' use of their property have often turned into contentious political issues. Such measures—usually to protect wetlands, coastal areas, or endangered species—have provided a new generation of debate about one permanent tension in American political life: the conflict between public interests in how land is used and the interests of landowners in deciding for themselves what to do with their own property. They have produced constitutional challenges, prompted legislative efforts to limit government authority, and provided ammunition for partisan campaigns.

Recent constitutional challenges have focused on circumstances in which the government must pay landowners for reducing their property's value through environmental regulation. The Fifth Amendment to the Constitution provides that private property shall not "be taken for public use, without just compensation," a provision that applies to state action as well as to federal action by means of its incorporation into the Fourteenth Amendment. The two extremes have long been clear. A government may use its power of eminent domain to take title to property for public uses such as highways or parks, compensating the owner for its full value. At the other end of the spectrum, a government may regulate the use of private property in ways that impair its value to some degree without compensating the owner because, in the words of Justice Oliver Wendell Holmes, "Government hardly could go on if . . . values . . . could not be diminished without paying for every such change."[21] Of course, some government actions also increase land values.

Since the 1970s landowner suits demanding "just compensation" for property value losses due to environmental regulations have reawakened an issue that was contentious in the early 1900s, when rapid industrialization created conflicts among business, agricultural, and residential land uses. In 1926 the conflict between public needs and landowners' interests was resolved by the Supreme Court's approval of local government zoning authority.[22] In 1978 the Court upheld an action under New York City's historic preservation law that denied the owner of Grand Central Station the ability to construct an office complex in air rights over the station. Since then the Court has suggested that the issue of compensation turns on the owner's investment-backed expectations and the character and economic impact of the government's action.[23] In 1992 the Court confirmed that a developer who was prevented from building anything on his land because of a coastal regulation would be entitled to compensation.[24] And in two cases in 1987 and 1994, the Court emphasized that government restrictions had to show an essential nexus with a legitimate state interest.[25] The Court's inability to devise a test that would allow landowners and regulators to predict when an environmental rule would require compensation has suggested to some the need for legislative action.

Widespread support in the 1990s for legislative proposals to pay landowners for the loss of property value caused by environmental regulations indicates that such issues tap into core concerns of many voters. Those proposals cannot be ignored as simply part of a crusade by extreme property-rights groups. In 1995 the House of Representatives voted by a wide margin (277 to 141) to require that the government pay landowners when government action to protect endangered species or to control water pollution reduced land value 20 percent or more and required it to purchase land when such action reduced its value 50 percent or more. Robert Dole, Republican senator from Kansas and a 1996 presidential candidate, got 31 cosponsors for a companion bill in the Senate that never came to a vote.[26] Between 1990 and 1995, 26 state legislatures also approved various property rights measures, many of them compensating landowners for reduced property values due to regulation (though some states later modified or repealed the laws). Although the practical effects and permanence of such laws are not yet clear, compensation requirements have the potential to increase greatly the cost of federal regulation.[27]

Such measures may reflect opposition not to integrating environmental values into land use decisions but to the *way* in which such restrictions have been imposed. Surprise often has played a critical part in creating particularly contentious issues, fueling legislative and judicial challenges. From the regulators' point of view, quick action is needed to respond to congressional directives. From the landowners' point of view, sudden environmental restrictions change the rules in the middle of the game. In one case that reached the Supreme Court, a developer who bought coastal land in South Carolina to build homes in 1986 was stopped by the enactment of a state antierosion law two years later. As a practical matter, environmental priorities are slowly working their way into routine planning decisions. In many cities and towns, some environmental concerns have been integrated into local land use rules to some degree. And there is no sign that Congress is planning to repeal wetlands, coastal protection, or endangered species laws.[28]

Questions about how public responses relate to the timing, justification, and perceived benefits of environmental restrictions become more urgent when regulators tackle widely dispersed sources of pollution and conservation opportunities on large tracts of private land. The Rivers Protection Act approved in Massachusetts in 1996 illustrates the clash between such efforts and landowners' interests. To control runoff that caused most of the state's water pollution and threatened drinking water, the act prohibited most development within 200 feet of rivers. Although real estate and developer interests blocked passage of the law for seven years, more than 100 citizen groups fought for it. When it was finally approved, Republican governor William Weld jumped into the Charles River to celebrate the occasion. How the law will fare when local conservation commissions start ruling on building requests remains to be seen.[29]

Attempts by Congress to require changes in people's everyday habits also have been problematic. Government efforts to persuade the auto industry to adopt pollution control technology have succeeded. Government efforts to persuade drivers to change their habits generally have not. Public reaction against measures such as restrictions on downtown access and parking that were part of federally mandated transportation control plans led Congress and the EPA to give up on such plans. In the 1990s public reaction against renewed federal efforts to change driving behavior—this time by forcing employers in 10 urban areas with the worst air pollution to make changes in employee commuting patterns—led Congress to give up on that requirement in 1995. And car owners' complaints about stepped-up federal requirements for pollution inspection in heavily polluted areas have led Maine, Pennsylvania, Texas, and other states to suspend such programs.[30]

Obviously, people do sometimes change their ways in response to government requirements, and more research would be helpful to suggest why such requirements succeed or fail. For example, many people have been willing to recycle cans, bottles, and newspapers. This change of habit is cheap, convenient, and may be perceived as directly reducing waste. According to Ladd and Bowman, 63 percent of people polled said they were at least occasionally separating trash for recycling in 1990, compared with 49 percent in 1980.[31]

Increasing constraints on public spending and rapidly expanding government responsibilities have raised the specter of hollow government. Federal and state laws add ever-increasing layers of requirements that may demand new levels of environmental protection no one is willing to pay for. The federal government's declining share of public spending for environmental protection and taxpayer revolts that have effectively capped spending in some states threaten to make rapidly increasing government responsibilities for pollution control and conservation a hollow shell.

In public opinion polls, people consistently say that the country is spending too little on improving the environment. But when asked about specific dollar amounts, most are willing to spend little in increased taxes and fees and expect companies to improve pollution control without increasing the price of products. "Incomes are modest," Ladd and Bowman conclude. "Many feel strapped, and there are so many problems to solve."[32]

Yet the public has in fact supported an enormous indirect investment in environmental protection since 1970. Business makes most of that investment (financed by consumers, shareholders, or employees) by buying new equipment and revising production techniques, for example. Local government pays the second largest share, mainly for improving public infrastructure such as sewage treatment, drinking water supply, and trash disposal. But voters (or their elected representatives) have skimped on the small portion of costs that supports government's efforts to carry out legislative mandates. That is a critical problem because national laws often make government the gatekeeper for new economic activity that increases pollution. When the line lengthens in front of the gatekeeper, everyone loses.[33]

As environmental protection has matured as a policy issue, organizations that promote or oppose specific measures have continued to use rhetoric that polarizes debate. But, in fact, times have changed. Many environmental groups also employ pragmatic approaches and join in ad hoc alliances with businesses. Where they once united to oppose large corporations and to change the political system, several groups have supported the use of market incentives or have formed alliances with big business. In 1990 the Environmental Defense Fund split from other environmental groups by supporting emissions trading as a new approach to reducing acid rain in the Clean Air Act amendments.[34] The Nature Conservancy has fought for tax incentives to encourage landowners to create conservation easements.[35] And a number of environmental groups, the Environmental Defense Fund and the World Wildlife Fund among them, have worked with large corporations to improve their environmental practices. In the 1990s the Environmental Defense Fund formed the Alliance for Environmental Innovation, which concluded agreements to improve packaging and make other changes with such companies as McDonald's and United Parcel Service.[36]

Some environmental groups have gone a step further and engaged in business activities themselves. Conservation groups have bid against ranchers to lease many thousands of acres of public land in western states.[37] Douglas Foy, head of Boston's Conservation Law Foundation, formed a for-profit insurance agency in 1998 to offer discounted policies to drivers who joined a company affiliated with the foundation, with incentives for limiting driving miles. The year before, Foy tried to buy New England Electric System's power plants in order to close the five dirtiest plants.[38]

Environmental groups are also divided on fundamental issues. In 1998 the membership of the Sierra Club was split on the issue of limiting immigration to stem future population increases, a position opposed by environmental justice groups.[39] The debate over the approval of the North American Free Trade Agreement (NAFTA) signed by Mexico, Canada, and the United States in 1993 drove a wedge between groups that are often allies. The Friends of the Earth, Public Citizen, and the Sierra Club criticized the treaty for being weak on environmental protection, while the Natural Resources Defense Council, the Audubon Society, and the Environmental Defense Fund, which helped negotiate accompanying environmental provisions, supported it.[40]

Federal Capabilities Grow While Influence Shrinks

In environmental protection, the overriding theme of the past 30 years has been a successful bipartisan campaign to increase the reach and specificity of national laws. The growing reality, however, has been an apparent erosion in the federal government's ability to make those laws work. International political and economic forces, decreasing federal leverage over the state and local governments, a widening information gap, an increasingly unmanageable workload, and a legal framework sometimes out of step with scientific understanding have set de facto limits on national pollution control and conservation efforts.

In 1990 Republican president George Bush signed amendments to the Clean Air Act that introduced the most far-reaching air pollution controls in 20 years, setting up new programs for hazardous air pollutants, acid rain, and automobile controls. The law was supported by a large majority of Republicans and Democrats. It passed by a vote of 401 to 25 in the House of Representatives and 89 to 10 in the Senate, after months of negotiations led by then Senate Majority Leader George Mitchell (Democrat of Maine). In addition to increasing the breadth of air pollution controls, the law increased the specificity of legislative requirements. The 1990 law ran more than 300 pages and included 162 statutory deadlines. By contrast, the 1970 Clean Air Act that initiated national standards ran 38 pages, with 12 deadlines.[41]

Federal action has increased the reach of national environmental rules in three ways: by regulating more sources of pollution and threats to conservation, by controlling a progressively wider array of pollutants and activities, and by increasing the stringency of rules. Those trends continued in the 1990s, despite divisions between a Republican Congress and a Democratic president. The 1990 Clean Air Act amendments expanded the category of businesses required to adopt new technologies to control toxic air pollutants to include dry cleaners, commercial printers, and solvent cleaning and chrome plating operations, among others.[42] In 1997 the EPA added mining, utilities, and other industries to the list of businesses required to disclose to the public hazardous emissions under the Toxic Release Inventory.[43] The EPA also imposed much stricter standards for fine-particle and ozone-contributing pollutants under a court order enforcing Congress's requirement that national standards be revised to reflect advancing science. Amending the Safe Drinking Water Act in 1996, Congress added a degree of local flexibility but also added strict new reporting requirements for local governments.[44] Only rarely has Congress backtracked by eliminating an area of federal regulation, as it did in 1979 by ending funding for noise pollution control, the environmental impact of which is clearly local.[45]

Even as national mandates have expanded, the federal government is losing some of its control over environmental policy. Political choices are increasingly circumscribed by international political and economic forces over which the president and Congress have little control. In his 1999 budget, for example, President Bill Clinton proposed to invest $6.3 billion to help carry out an international agreement to reduce the risk of global warming, an environmental policy that the U.S. government did not initiate and spent years trying to avoid, aimed at controlling gases— mainly carbon dioxide—that have not been regulated nationally as pollutants.

The power of the federal government to make environmental policy is limited by an increasing number of multilateral agreements to address problems now recognized as essentially international in character, by growing international trade, and by a nascent recognition that the United States cannot permanently ignore environmental degradation in developing countries. By 1996 more than 200 international environmental agreements governed a broad range of environmental issues including pollution that crosses national boundaries, import and export of hazardous waste, trade in endangered species, commercial fishing, the protection of wetlands, ozone depletion, and (if Congress approves) global warming.[46]

To cite a particularly successful example of international action that limited national choices, scientific evidence that common chemicals used in refrigeration were contributing to the thinning of the earth's protective ozone layer led to multilateral agreements that altered industrial processes and consumer products in the United States. Under agreements reached in 1987 and 1990, the United States and other countries that were major producers of ozone-depleting compounds known as chlorofluorocarbons pledged to phase out the main chemicals by the year 2000.[47]

With the end of the cold war and a steady decline in transportation and communication costs, trade in goods and services and the international movement of capital have expanded enormously. As tariffs are reduced or eliminated, multilateral agreements that have accompanied these changes are increasing the pressure for lower nontariff barriers to trade as well. From a free-trade standpoint, environmental rules can amount to such a barrier, by limiting imports of products that do not meet national standards. Under the General Agreement on Tariffs and Trade (GATT) and NAFTA, foreign countries have charged that rules such as auto fuel–economy standards or fishing requirements to protect dolphins and sea turtles can amount to unfair trade barriers when applied to imported products. For example, in 1998 the appellate body of the World Trade Organization ruled that a U.S. attempt to import shrimp only from countries using practices to protect sea turtles was an arbitrary and unjustifiable discrimination in trade. In 1991 a GATT dispute settlement panel ruled that a U.S. import ban on tuna caught in ways that harmed dolphin was a violation of the trade agreement.[48] And the increase in trade itself also means that products made in the United States increasingly compete with products manufactured around the world, thereby creating pressure for greater similarity in environmental standards. U.S. environmental regulators cannot afford to ignore standards imposed by other countries when profits of American firms are affected by differences between those standards and U.S. rules.[49]

Finally, Congress finds it increasingly difficult to make national policy that does not take into account severe pollution and conservation problems in developing countries, for three reasons. First, some links between such problems and U.S. national security interests are now widely acknowledged. Extreme environmental degradation can erode economic and political stability, and incidents such as the radiation leak from Russia's Chernobyl nuclear power plant in 1986 have helped highlight such connections. Second, improving science has underscored the international character of issues such as biodiversity and global warming. Third, advances in computer and communications technology continue to heighten public awareness of human suffering and other effects of such degradation.

As a practical matter, national leverage over state environmental actions also has weakened, even as federal authority has increased, owing to three converging trends. First, a 20-year decline in the share of environmental protection funding provided from the national budget has limited the influence of federal grant conditions on state action. Federal grants for pollution control peaked in the late 1970s, and shrank 21 percent between 1980 and 1994.[50] The largest national environmental program, a 20-year effort to upgrade local sewage treatment plants with $66 billion in federal funds, was scaled back and converted to a revolving loan fund in 1987.[51]

The federal share of environmental funding varies widely from state to state, and reliable information on exact proportions is lacking. The Council of State Governments estimates that the federal contribution now averages 26 percent of the $10.7 billion that states spend on pollution control and conservation, but EPA staff suggest that the federal share may range from as high as 50 percent in states with limited state programs to as low as 8 or 9 percent in states with extensive programs.[52]

A second trend is the federal government's propensity to place increasing constraints on its own authority. Both Congress and the executive branch have erected new obstacles to the creation of federal directives that are not accompanied by resources to carry them out, for example. In October 1993 President Clinton forbade federal agencies to place new "unfunded mandates" on state and local governments without consulting with them. In March 1995 Congress requested detailed cost estimates for unfunded requirements and provided that those costing $50 million or more could be stricken from bills on a point of order. (The law did not apply to administrative actions under existing laws.)[53]

A third trend is evident in several recent, closely decided cases in which the Supreme Court has begun to define some limits to Congress's constitutional authority to direct state action. In 1992 the Court ruled that the federal government could not force states to take ownership of low-level radioactive waste.[54] In 1995 it ruled 5 to 4 that Congress's reach to regulate interstate commerce does not extend to prohibiting possession of firearms near schools.[55] In 1997 it struck down by a vote of 6 to 3 a federal law that restricted state authority to apply zoning regulations where doing so would affect religious practices.[56] Equally significant, as state capabilities continue to increase, they, too, place de facto limits on federal choices (as discussed later in the chapter).

Practical limitations also have prevented the federal government from doing some of the jobs it is supposed to do best. Persistent gaps in fundamental information to support government and business decisions, weak efforts to control interstate pollution, an overwhelming administrative workload, and an outdated framework of laws reduce federal effectiveness.

Political sensitivities have sometimes constrained federal support of environmental research. Congress has repeatedly turned down proposals to gather information about species of plants and animals that would make it possible to identify ecologically critical lands, bowing to groups arguing that more information would lead to more regulation of the use of private property. As a result, transportation planners, developers, and landowners often rely on conservation information collected by private groups such as the Nature Conservancy. The Conservancy's Natural Heritage programs, administered mainly by state governments, are databases that aim to track the location of rare species.[57] By sometimes assigning a low priority to environmental research, politics perpetuates gaps both in fundamental understanding of pollution and conservation processes and in monitoring capabilities, thereby placing practical limits on progress. Even when research is conducted, its findings are sometimes ignored in political decisions. A 10-year study of the causes and effects of acid rain that cost half a billion dollars was given little attention by Congress in its deliberations on amendments to the Clean Air Act in 1990.[58] (Information issues are discussed further in chapter 4 and the epilogue.)

Ironically, Congress and the executive branch for many years chose to play a weak role in curbing interstate pollution as well. Under federal laws, state air pollution control plans were supposed to address interstate pollution, but such provisions were narrowly applied. In the 1970s, with a strong federal air pollution law on the books, power companies built scores of smokestacks over 500 feet high to carry pollution away from local areas. Initially, the EPA interpreted congressional requirements that state plans minimize interstate pollution as mainly a requirement for the exchange of information. In the 1980s uncertain science concerning long-distance transport and chemical changes in pollution reinforced political inclinations to limit national action. It was not until 1990 that Congress tackled the problem of interstate air pollution and its contribution to acid rain in earnest, by instituting a trading program for sulfur emissions. And only in 1998 did the federal government adopt a multistate plan to reduce nitrogen oxides that drift from midwestern power plants to eastern cities.[59]

In the late 1990s lawsuits by one jurisdiction against another continued to consume valuable time and resources. In March 1998, for example, New York State and Connecticut filed separate suits against New York City, charging that it had solved one environmental problem by creating another. Barred by federal law from dumping sewage sludge into the ocean, the city squeezed it dry in order to transport it to landfills. But suits charged that that process produced nitrogen-rich wastewater that polluted Long Island Sound.[60]

The federal government also has been slow to tackle the huge task of cleaning up pollution caused by the military. The total cost of cleaning up pollution on government property from military programs may be as high as $389 billion ($350 billion of it related to nuclear weapons programs and $15 billion of it related to cleaning up and returning to civilian use eight chemical weapons sites). Nuclear cleanup includes 115 facilities (15 of them major operations) in 34 states and territories, and more than 7,000 locations where radioactive or hazardous substances were released into the environment. Other main military contaminants include fuel, solvents, and corrosives at 10,000 sites where chemicals were inadequately disposed of, leaked, or spilled, and environmental damage occurred. Congress told the Defense Department in 1984 to evaluate and clean up these wastes. Twelve years later, the General Accounting Office found that the military had evaluated only 70 percent of the sites and "does not have sufficient data to manage its environmental compliance programs." The president's budget for fiscal 2000 proposed $10.2 billion for ongoing military and nuclear cleanup, an amount larger than the EPA's total budget.[61]

The federal government's inability to keep up with the huge regulatory workload created by environmental laws also acts as a de facto limit on national authority. Timely government action is important because many pollution and conservation laws are constructed around agency approval of each major activity with environmental consequences. As a result, government inaction or delayed action not only increases controversy but also perpetuates uncertainty and can block both economic and environmental progress.

The incompatibility of two enduring congressional goals—quick action and ample opportunity for public participation—contributes to this problem. Most federal laws require both and make no attempt to reconcile the two. The 1990 Clean Air Act, for example, required the EPA to produce 100 new rules in two years, including regulations for a new permit program for existing factories but also set new requirements for public participation. Such participation nearly always lengthens the time taken to reach a final decision.[62]

Finally, the framework of laws constructed in the 1970s has itself become an obstacle to effective federal action. At least 11 cabinet agencies in addition to the EPA share responsibility for environmental protection. Separate staffs within the EPA deal with air, water, and waste problems, though the agency is making an effort to improve connections among those programs. Those compartments do not reflect evolving knowledge. Scientists and managers have long understood that air, water, and land pollution are interrelated, that conservation and pollution control are not separate concerns, and that limiting discharges often is less effective than encouraging production processes that prevent pollution.[63]

During the 1990s the EPA and the Department of the Interior have attempted to reform the reforms of the 1970s to reduce the political tension between new needs and old laws, without waiting for Congress to act. They have taken tentative steps away from uniform standards toward industry-by-industry or case-by-case rules, encouraged negotiated agreements, launched pilot projects to allow businesses more flexibility in pollution control in return for superior environmental performance, and permitted individualized state plans (see chapter 4).

Whether these actions are the beginning of a transition to more effective national action or whether they are simply temporary coping strategies is not yet clear. Case-by-case approaches themselves create new demands on tight budgets, and at the EPA much progress has occurred "only at the margins," according to former administrator William D. Ruckelshaus.[64] Such measures do, however, underscore an irony of the changing times. Laws that were intended to limit administrative discretion during a time of congressional activism have become a vehicle for administrative experiment during a time of congressional gridlock.

States "Race to the Bottom" or Race to the Bottom Line?

In 1998 Oklahoma's Republican governor, Frank Keating, announced strict new environmental controls on one of the state's leading industries. He issued executive orders and proposed new laws to require state licensing of large hog and poultry farms, and to demand soil testing and groundwater protection. The governor explained that "other states have discovered the hard way that the issue of water quality must not be ignored. Reasonable restrictions must be placed on the industry now to preserve the water supply for the future."[65] He was referring to incidents such as a 1995 spill in North Carolina that dumped millions of gallons of liquid hog manure from a clay-lined lagoon into nearby rivers, inspiring a two-year moratorium on new hog farms, the state's leading agricultural business.[66]

Stories like these underscore the need for new thinking about state interests in environmental protection. Thirty years ago state recalcitrance was one impetus for national action. Now, a widely perceived need to adjust state flexibility in carrying out national laws means revisiting troublesome questions: to what degree can states be trusted to control pollution and improve conservation? When is it all right for each state to decide on its own standards?

The assumption that states engage in a "race to the bottom" to minimize environmental protection in order to attract business is too simplistic a notion for the 1990s. The idea is outdated for three reasons. First, empirical evidence is by now overwhelming that businesses rarely decide where to locate or expand on the basis of the strength or weakness of state environmental programs. Second, state politics has been transformed in ways that make it more likely that pollution and conservation issues will have a prominent place on political agendas. Third, government officials, business executives, and the voting public are in-

creasingly pragmatic—they understand that some environmental measures contribute to state prosperity whereas others clearly do not.[67] The
challenge now is to sort federal and state responsibilities to take advantage of the strengths of each level of government.

Environmental costs seldom are a determining factor in the location of
a business because they are a small and usually predictable element of
business operating expenses, dwarfed by labor, real estate, and transportation costs in relocation decisions. Even for chemical and petroleum industries, annual pollution abatement expenses run less than 2
percent of sales. Capital expenditures for pollution abatement, which
average about 7.5 percent of total capital costs, vary widely among industries, ranging from 2 percent of total capital costs for machinery and
3 percent for electric and electronics to 13 percent for chemical industries and 25 percent for petroleum and coal. Even when substantial,
though, those costs usually are less influential in relocation decisions
than labor, real estate, transportation, energy, and tax considerations,
according to surveys of corporate executives and relocation experts.[68]

Also, relocations are relatively rare in most industries, and evidence
indicates that business moves may be a small part of a state's economic
health. According to the Council of State Governments, most studies
show that in the past 10 years, 8 out of 10 jobs have been created locally,
mostly in connection with the expansion of existing small and middle-
sized businesses.[69]

Empirical evidence confirms that the rigor of state environmental
policies generally does not have much influence on where businesses
move. Economists have found the issue hard to analyze because of the
paucity of information about business relocation and the complexity of
environmental policy, but in general they have found no strong association between environmental compliance costs and business moves.[70]
Likewise, there is little evidence that international businesses seek pollution havens, according to the Organization for Economic Cooperation and Development (OECD).[71]

The modest overall impact of environmental costs does not deny,
however, that changes in pollution rules can close down factories and
destroy jobs on occasion, especially if they are sudden and unexpected.
Retrofitting old factories to satisfy new requirements can be extremely
expensive, and small or marginal businesses cannot always survive
government demands to make unexpected changes. Likewise, the ab-

sence of evidence that businesses relocate overseas to avoid environmental regulation does not deny the possibility that U.S. rules could get so far out of line with those of some other countries that they might become an important consideration in the relocation of a business. That is unlikely, however, when the worldwide trend is toward strengthening such rules. In the end, of course, what influences policy most is the *beliefs* of governors and legislators about why businesses relocate rather than the weight of empirical evidence.[72]

The states of the 1990s are very different from the states of the 1960s, and their role in environmental protection has expanded enormously. As political entities, they have been transformed by consolidation of authority in governors' offices, by reapportionment of state legislatures, by the growth of professional staffs, by increased competition between political parties, and by reforms that ensure greater public participation in regulatory decisions.[73]

Many aspects of environmental protection have been assimilated into state and local politics, as they have been into national politics, thanks in large part to 30 years of national requirements. State programs sometimes are broader than federal requirements, and federal funding now accounts for less than a third of total state environmental funding, according to the Council of State Governments. In most states, single agencies manage pollution control (and sometimes conservation as well) with increased expertise, and in 1994 the heads of those agencies formed the Environmental Council of the States to encourage state initiative and to lobby for federal flexibility.[74]

Echoing the National Environmental Policy Act, most states require an assessment of environmental impact of important public actions, and a combination of state and federal rules provide many opportunities for citizens to challenge government action or inaction. Among them are public notice and comment before regulations are finalized; public hearings before permits are granted to industry to discharge pollutants in the air, in the water, or on land; and the right to challenge in court whether state agency actions are reasonable measures under state and federal pollution laws.[75]

State and local governments are responsible for carrying out nearly all national environmental laws and have continued to dominate decisions in matters such as land use and waste disposal where congressional action is limited. By 1994 the federal government had granted to

38 states authority to issue and enforce water pollution permits, to 46 states authority to carry out national hazardous and solid waste requirements, and to 39 states full authority to carry out air pollution requirements for new sources.[76]

As a result of these changes, companies that seek to relocate may get an unexpected dose of state scrutiny of their environmental practices as part of the bargain, even in a state that is competing hard to gain business. When Intel, Inc., producer of the pentium computer chip, announced a decision in 1993 to build the world's most advanced semiconductor plant in Rio Rancho, New Mexico, local protests directed national attention to pollution and conservation issues in an industry generally seen as environmentally friendly. Opposition induced Intel to cut proposed water use and to spend unanticipated millions on pollution control.[77]

Changes in state politics also increase the chances that lax enforcement will be self-correcting. When the EPA sued Smithfield Foods, Inc., in 1996 for dumping hog wastes into the Chesapeake Bay and criticized Virginia's governor, George Allen, for a three-year pattern of failing to enforce water pollution laws, the regional Chesapeake Bay Foundation had already called attention to violations and a bipartisan panel of the state legislature had issued a 255-page report attacking the laxity of that state's enforcement program.[78] When James S. Gilmore III replaced Allen as governor in 1998, he promised to protect the state's natural resources as one of his top three priorities, along with controlling crime and building an economy based on information technology.[79]

The environment does not always win, of course, and business lobbying remains powerful, but pollution control and conservation now have a permanent place on state and local, as well as federal, political agendas, where they compete with varying degrees of success with other pressing priorities. To the federal floor of national standards, governors and legislators have added a variety of independently supported state structures that outlast changes in political regimes and that act as a counterweight to political moves to cut environmental protection.

Regional initiative also is becoming more common, often promoted by federal actions. States that share a common watershed have begun to identify sources of pollution and discuss controls, a process encouraged by the EPA since at least 1994.[80] Starting in 1995, 37 states and the District of Columbia worked for two years under the auspices of the EPA to find ways to reduce the drift of smog-causing emissions of nitrogen oxides over the eastern United States. The group remained divided between upwind midwestern states where many older coal-burning power plants are located, and downwind Eastern states where pollutants are blown. But the effort helped expand scientific knowledge, and a majority of states recommended reducing smog-related pollution from power plants by as much as 85 percent and pollution from factories by more than two-thirds, a plan that had to be approved and put into action by the EPA. In September 1998 the EPA announced a plan to reduce such emissions by an average of 28 percent in 22 states by 2003. The federal plan left states free to decide how reductions would be achieved, but encouraged the establishment of an emissions-trading system similar to the one in operation to reduce sulfur dioxide emissions associated with acid rain.[81]

Sometimes, regional cooperation can spread the costs of addressing a complex problem or concentrate pressures for action among affected states. But such cooperation has its limits: it is unlikely to alter fundamental state interests. National legislation aimed at getting states to work together in deciding where to dispose of low-level radioactive wastes (meaning all commercial and medical waste except waste from nuclear power plants) produced only two commercial sites by 1996. Low-level radioactive waste accounts for 85 percent of radioactive waste in the United States, though less than one-tenth of a percent of radioactivity in waste. States have formed interstate compacts, as intended by the legislation. But choosing permanent sites has remained an explosive issue. The result is that waste is stored mainly at scattered, temporary sites around the country.[82]

The race among states today is a competition to gain prosperity in a fast-changing economy in which firms are consolidating, capital is becoming increasingly mobile, and skilled workers are sometimes in short supply. The contest is not a "race to the bottom" to minimize environmental protection, but a "race to the bottom line" to improve property values and increase tax revenues.[83] In these tumultuous times, it is important to understand as much as possible about states' environmental interests in order to promote effective policies.

When states compete to gain prosperity, environmental protection plays an interesting and complex role. Two overarching themes are that some environmental measures help states compete, whereas others plainly do not; and, at least at the extremes, states with strong economies tend to support strong environmental protection programs, while those with weak economies often support weaker programs.

From a state government viewpoint, proposals to improve environmental protection may be seen as desirable investments or burdensome liabilities.[84] As a practical matter, at least five factors may influence a state's decisions about whether a particular environmental measure is a good investment. The first and most obvious is the degree to which it is clear that the government has the capacity to act. Capacity includes basic questions such as whether there is legal authority to regulate and whether a feasible approach is available. Another factor is whether benefits are expected to be concentrated in the state. Residents are likely to be more inclined to spend money or make sacrifices if they can be reasonably sure that their state will enjoy the rewards from them. Two other considerations are whether improvements are directly related to human health or to ecology with human consequences, and whether they are likely to be experienced soon. Political pressures that emphasize actions with clear links to results residents can experience and with short time horizons for benefits are particularly strong at the state level of government (see chapter 4). A final factor is whether action contributes to a jurisdiction's current vision of how to achieve prosperity. A state's more or less cohesive idea of how to achieve prosperity is likely to be influenced by current economic and political forces as well as by the state's history and culture, is likely to change as those forces evolve, and helps to make each state unique within the federal system.[85]

Some examples of state-initiated investments in environmental protection may help to clarify how such factors work. Long before federal regulation, states and communities set up systems to provide safe drinking water, remove trash, and promote other public environmental infrastructure that prevented epidemics and improved the jurisdiction's reputation. More recently, states have found it politically palatable to regulate leading businesses when environmental benefits are direct and immediate. In the 1990s several states placed moratoria on large new animal-feeding operations and began regulating existing operations when they discovered that large manure lagoons posed a threat to water quality. Federal regulation followed in 1998, requiring permits for the largest farms and encouraging voluntary compliance by smaller operations.[86]

Some states and communities have linked future prosperity to programs designed to control residential growth, both to avoid infrastructure costs and to conserve remaining open space. In 1997, for example, predictions by Governor Parris N. Glendening that development in the central part of the state would gobble up as much land in the next 25 years as it had in the state's entire history helped persuade Maryland's state legislature to approve a particularly strong growth management initiative. Using the state's $15 billion budget as an incentive, the law encouraged the application of state funds for new roads, sewers, and schools in areas targeted for concentrated growth, "to save our natural resources before they are lost forever," Glendening said. The law also allotted $140 million to acquire land threatened by development.[87]

New Jersey's Republican governor Christine Todd Whitman also announced a $1 billion growth-control program in 1998. Supported by an unusual coalition of 10 major corporations—including Mobil Oil, Dupont, and Bell Atlantic—and by 10 environmental groups, the fiscally conservative governor proposed raising the state gas tax to finance the purchase of nearly half the undeveloped land in the state to save it from development. The state legislature endorsed the $1 billion plan (funded from borrowing and present taxes) and voted to put the issue on the ballot, where it was approved.[88]

In the November 1998 elections, voters in more than 200 jurisdictions cast ballots on growth management initiatives, and most were approved.[89] Federal proposals followed. The Clinton administration highlighted the issue of planned growth in its fiscal 2000 budget, proposing $900 million in spending from the federal Land and Water Conservation Fund and $9.5 billion from Better America Bonds, a new financing mechanism to support state and local conservation efforts.[90]

None of these developments change the fact that there are difficult trade-offs between environmental protection and economic development. Conflicts between the two can be politically divisive, especially when quick changes are expected. Hiring inspectors to enforce pollution laws or buying land to protect a watershed is expensive and must vie for limited state funds with initiatives directed at improving schools, building roads, and paying for medicaid and welfare. Nor do these developments mean that environmental interests will prevail. When issues pit jobs against cleaner air or conservation, business, labor, homeowners, and other groups fight to protect their interests.

Also, business influence in state and local politics should not be discounted. Business interests are powerful, and their importance in state capitals may be growing as some lobbying efforts move from Washington to the states. But these days business interests are rarely united on environmental issues. Some companies usually stand to lose while others tend to gain from particular environmental regulations, as discussed in the next section. Furthermore, to say that business competes with other interest groups is quite different from saying that there is an inevitable economic imperative that causes states to minimize pollution control and conservation in a "race to the bottom."

Finally, the idea that states find it in their interest to promote some aspects of environmental protection should not be taken as an argument that the federal government should get out of the business of pollution control. To emphasize an earlier point, most environmental issues on which Congress has taken action include a mix of national, state, and local interests. The tough questions concern how, not whether, responsibilities should be shared.

At the other extreme, a state is likely to avoid taking action to improve environmental protection or is likely to take action that is more permissive than prescriptive if it lacks the legal and practical capacity to tackle a problem; if benefits are not concentrated in the state; if improvements are not directly related to human health or to ecology with human consequences, or will not be experienced soon; or if action does little to promote the state's current vision of how to achieve prosperity. Spending money to clean up pollution that drifts, flows, seeps, or can be transported to other states is likely to be viewed as a poor prospect by state politicians. Much of the detritus of civilization turns out to be mobile. The march of science has revealed that some air and water pollution can travel great distances and can change form along the way. Scientists rate toxic air pollution from distant sources as a primary cause of water pollution in the Great Lakes, link sulfur dioxide emissions from midwestern power plants to acid rain in the northeastern states and Canada, and trace nitrogen oxide emissions from distant factories to East Coast smog.[91]

Sewage treatment is a good example of a state liability that has caused many political disputes. Benefits of treatment are often indirect and accrue to residents of distant communities or states as treated water flows downstream. Early cities were located on lakes or rivers because water was a free transportation system for goods and people. But water was also a terrific transportation system for sewage and garbage. The history of American cities around the turn of the century is riddled with lawsuits brought by one city or state against another for sending pollution downstream.

Around 1900 Chicago went further and undertook what was then thought to be the world's largest public works project when it decided to reverse the flow of the Chicago River in order to send 1,500 tons of daily sewage to St. Louis instead of to Lake Michigan, its drinking water source. Missouri charged that the sewage was a public nuisance and that it was responsible for a 77 percent increase in typhoid cases. In his Supreme Court opinion, Justice Oliver Wendell Holmes expressed doubts about the data but also ruled that Chicago's action was not a public nuisance because it was general practice. St. Louis, it turned out, also sent its sewage downstream, as did most other cities.[92]

Governors and legislators also are unlikely to see economic or political benefits in giving up prime development land to promote biodiversity or to protect endangered species. Environmental scientists must consider the effects of development on future generations, and national environmental groups keep concerns about biodiversity and endangered species before the members of Congress. But the unavoidable necessity of competing with their rivals keeps states focused on short-term gains. When liabilities for individual states are priorities for the nation, rigorous federal oversight is needed.

As states gain in capability and are charged with mounting monitoring and enforcement tasks, trends in federal and state spending raise the specter of increasing environmental responsibilities that no level of government is willing to pay for. The threat of hollow government is more severe for states than for the federal government, because states grant most permits and take most enforcement actions. Monitoring and enforcement expenses are a tiny share of national environmental costs, most of which are borne by industry. But their adequate funding is critical in a system in which government is the gatekeeper for new economic activity that causes pollution or influences conservation.[93]

Evidence is accumulating that state and local governments are unable to muster the resources to carry out even the main provisions of national pollution control laws, despite periodic budget surpluses. State capabilities keep increasing, but regulatory burdens increase faster. "The costs of implementing the growing number of environmental regulations mandated by Congress are overwhelming the budgets of many state governments," the General Accounting Office (GAO) found in 1995 after surveying 16 states and examining its previous studies. Nearly all of these states said that they did not have the resources to carry out even the main provisions of national laws and could not accept more responsibilities for the control of drinking water, solid waste, or water pollution without cutting back further on key activities, nor could they come up with the funds needed just to keep pace with existing requirements. State governments appropriated only about half the resources needed to carry out federal drinking water requirements, and they allocated only 60 percent of the resources required to carry out water pollution control responsibilities. Many demands ultimately fall on local governments, where the funding squeeze is especially severe. Congressional action in 1995 that limited federal "unfunded mandates" to state and local governments did not stop the enactment of amendments to the Safe Drinking Water Act a year later that imposed large new costs on state and local governments. Congress chose to ignore its own recently enacted warnings.[94]

In such situations, the federal safety net, too, may be illusory. When interviewed by the GAO, EPA officials said that they, too, lacked the resources to meet their own performance criteria if they withdrew delegations from the states. In southern states, where the EPA operates some water pollution control programs directly, the GAO found that more than 5,000 facilities were operating with expired permits or with no permits owing to a shortage of government resources and industry neglect. The federal government spends about twice the $10 billion that states allot to environmental protection annually, but that includes substantial amounts for military cleanups.[95]

For states, the funding squeeze is the result of two gradual changes over the past 20 years. Starting in the late 1970s, Congress reduced the federal government's share of funding for major domestic programs, including environmental protection. That trend was reinforced by a Reagan administration strategy of weakening national environmental programs by cutting their funding. At about the same time, voters began to rebel against attempts to raise state taxes, a trend that continued in the 1990s. That taxpayers' revolt was symbolized by California's Proposition 13, a ballot initiative approved by voters in 1978 that slashed state property tax revenues by 57 percent by limiting property taxes to 1 percent of property value and that required a two-thirds vote of the state legislature for new taxes. Pressure for more services for growing numbers of retirees and children has added to the severity of the funding squeeze. National environmental laws also have increased the cost of local infrastructure such as sewage treatment plants and landfills.[96]

State and local governments have always ridden what Alice Rivlin has called a "fiscal roller coaster," raising or lowering taxes and service levels as economic times change to meet state constitutional requirements that budgets balance.[97] In good times, growing revenues from economic growth sometimes have allowed state governments to increase spending and hire new employees without raising taxes. In the mid-1990s, for example, states increased spending modestly to meet education, corrections, and other pressing needs, squirreled away some revenues in rainy-day funds, and cut taxes.[98] But in bad times, as in the recession of the early 1990s, the funding squeeze becomes severe as Congress cuts spending and voters continue to limit state taxes.[99]

Like the federal government, state governments are subject to practical pressures that reduce their effectiveness at a time when responsibilities for environmental protection are increasing. The growing complexity of national and state rules, growing opportunities to challenge them, and increased competition between political parties make it harder to reach lasting decisions. In 1997, 30 states had divided governments, meaning that the governor was of a different party than at least one house of the legislature.[100]

State authority is also challenged by the march of science, which reveals that problems once considered local may have distant effects. Scientists now agree that water pollution often originates as air or land pollution from neighboring states. An estimated quarter of the nitrogen pollution in the Chesapeake Bay, for instance, comes from air pollution. Nearly all the PCBs in the Great Lakes are deposited from the air.

Waste disposal is arguably an inherently local problem, though it is currently nationally regulated. Considered arguments have been made in favor of placing landfill liners and other specifications under state or local jurisdiction.[101] But any such reconsideration would have to contend with the fact that waste disposal practices can also have national and international consequences. Pollution from landfills can seep into ground or surface water that moves from state to state. Rotting garbage produces large quantities of carbon dioxide and methane gases now linked to global warming. New York City's Fresh Kills landfill, the grandfather of all waste disposal sites, produces 30 million cubic feet of landfill gas and discharges one million gallons of polluted water each day, even though it has its own waste treatment plant. In addition, interstate shipment and disposal of waste is a big business, raising the prospect of complex litigation over differing state rules when accidents occur. And population growth makes some local problems into interstate issues. A third of the nation's metropolitan areas encompass more than one state, and that number is growing.[102]

Because experience has shown that wealthy economies tend to devote more resources to federal environmental priorities than do struggling ones, national policymakers should be especially concerned about the future of pollution control and conservation in less affluent states. Just as differences among states in public spending generally have stayed about the same for 25 years, so, too, have differences in spending on environment and natural resources, with some states spending more than twice as much as others, even when population or manufacturing differences are taken into account.[103]

However, terminology may be deceiving. Relatively poor states or communities may devote considerable resources to environmental measures that are directly related to human health: trash removal, rat control, sewage regulation, and reducing the threats to children from lead paint, for example. If such actions are not among federal priorities, they are no less environmental commitments.

Some research has suggested direct links between prosperity and the kind of environmental protection emphasized in federal laws, at least at the extremes. "Providing cleaner air, water, and land is expensive," a Bank of America report concluded in 1993. "The more wealth an economy generates, the more resources it can devote to environmental improvement. Citizens who live in areas with strong economies tend to demand more environmental protection than citizens in areas with weak economies."[104]

Such differences among states are not surprising. State boundaries were drawn by accidents of settlement and political compromise, not by a desire for equity. Those chance divisions have produced variations in natural resources, population characteristics, political culture, and history, differences that in general the nation celebrates. They have also produced variations in taxable assets. State environmental protection, which lies at the junction of economic forces, political will, and historical tradition, naturally reflects such enduring differences.

States that have been dependent on oil, timber, mining, or other natural resource industries face special problems in their attempts to improve environmental protection and to assemble the ingredients of lasting growth. The cyclical character of markets for oil, timber, and minerals may lead businesses and residents to minimize permanent investments in areas dependent on those industries, while damage to the natural environment sometimes makes them unattractive places to live. These days, businesses sometimes move to places where skilled workers want to live, instead of choosing low-cost locations and expecting workers to follow.[105]

States that are weak in both economic growth and environmental protection are particularly vulnerable to the funding squeeze discussed earlier in this chapter for three reasons. First, many of the least prosperous states depend heavily on federal funds to finance their environmental protection programs. For the foreseeable future, such funds will be severely limited by other demands on the national budget. Second, states with low personal income, sales, and property values gain less revenue from tax rates that are comparable to those of richer states. Third, state budgets are likely to be heavily burdened by demands from social programs such as welfare and medicaid in less prosperous states, where the poverty rate also tends to be high.[106]

None of this should be construed as an argument for economic determinism, however. State economies are constantly changing as markets change, and experience has shown that political will and fortuitous circumstances can overcome obstacles to growth. High-technology industries, not oil, are now the biggest employers in Texas.[107] Booming high-tech industries and tourism made the Rocky Mountain states the fastest-growing region in the country in the early 1990s. In Idaho, electronics manufacturers such as Micron and Hewlett Packard employ more workers than the traditionally dominant forest products industry.[108] Tourism is the fastest-growing industry in the Rocky Mountain states and the largest private employer in seven western states.[109] Ohio's economy, shaken from its steel, rubber, and glass manufacturing base by sudden oil price increases in the 1970s, has re-formed around new manufacturing enterprises using advanced technology and more skilled workers. And restaurants standing where steel mills once lined the Cuyahoga River serve Burning River Pale Ale, a nostalgic reference to the burning debris that became a symbol of industrial neglect in 1969.[110]

Business Interests Vary

Paradoxically, business acceptance of environmental protection requirements has grown as American industry as a whole has become economically less secure. In the 1970s, when economic growth seemed unstoppable, executives of auto companies, utilities, and many other large corporations warned that the high cost of federal pollution standards would have devastating consequences for business. Some dug in their heels, delaying compliance. Now, a fast-changing economy and increased international competition have shaken the economic underpinnings of many companies. Between 1970 and 1990, U.S. exports doubled and imports tripled as a portion of GNP, and U.S. investors increasingly participated in an international capital market. Yet big business is making a large and rapidly increasing investment in environmental protection, with relatively little complaint. In 1994 companies spent an estimated $76 billion on pollution abatement alone, an amount that increased at double the rate of general economic growth in the mid-1990s. Today it is the practices of smaller firms and farmers that are new concerns of government regulators.[111]

On the other hand, big businesses have little incentive to take action beyond compliance, have increased incentives to minimize costs as they compete internationally, and still represent a broad spectrum of initiative and compliance. The idea that big business is a united force on environmental issues is a myth. Pollution control and conservation issues pit competitors against one another, big businesses against small ones, existing businesses against those that want to move in, and firms with deep pockets against those with none.

The fact that the American public, along with voters in other developed countries, has demonstrated a lasting commitment to improving pollution control and conservation exerts a powerful influence on business decisions. Company choices are not walled off from public opinion. Businesses are rewarded with profits for responding with agility to changes in economic and political forces. Changes in markets and politics, in turn, respond to evolving public attitudes.

For big businesses that sell directly to consumers or have a corporate image to protect, goodwill has economic value. "Nearly all major industrial leaders know that environmentalism is here to stay," William

Ruckelshaus, former EPA administrator and chief executive officer of a Fortune-500 company, explained 25 years after the first Earth Day demonstrations, "and so firms wish to avoid charges that they are insensitive polluters, just as they wish to avoid defects in quality."[112] As a symbol, the 1984 toxic gas leak at Union Carbide's pesticide plant in Bhopal, India, which killed 2,800 people and injured 50,000, remains current.

Every step that extends the reach of national or state environmental policy is contentious because it changes corporate balance sheets and rearranges competitive advantages. Environmental regulation can reduce the value of a company's capital assets with the stroke of a pen, by forcing early replacement of equipment. But business executives, like voters in both parties, now accept the premise that federal and state governments should promote pollution control and conservation. At least as a tactical matter, corporate leaders less often make charges of business-destroying costs and more often raise issues of uncertain science. "If we've learned any lesson, it's that you have to engage the debate on a different basis than costs," John M. McManus, manager of environmental strategies for American Electric Power, told a reporter from the *New York Times*.[113]

When public pressure mounts, corporate executives sometimes take preemptive action. U.S. auto companies lobbied for stricter national emission controls in 1997, partly to end decades of efforts by northeastern states and California to require some zero-emission cars.[114] In 1996, 200 timber and paper companies controlling 52 million acres in the United States adopted a "sustainable forestry initiative" to protect water quality and habitats, hoping to forestall further government action in response to clear-cutting. The move split both timber interests and environmentalists. Ten members of the sponsoring American Forest and Paper Association resigned, and the national Nature Conservancy supported the plan while the National Wildlife Federation and the Sierra Club criticized it.[115]

Many people assume that market forces work against business improvements in environmental protection because such improvements increase costs—and often they do. In a time of increasing national and international competition, one might expect pollution control to get short shrift. But such ideas need to be qualified. Often, as industrial efficiency increases, pollution plunges. And improvements in environmental protection do not have the same effects on each company, and firms searching for competitive advantage sometimes find it in new initiatives. In February 1998, for example, U.S. auto makers made the astounding announcement that they would voluntarily reduce hydrocarbon emissions in new cars by 70 percent and nitrogen oxides by 50 percent in one year. That announcement helped head off future regulation, to be sure. But it also responded to plans by competing Japanese companies to introduce low-emission vehicles.[116]

Manufacturers already using clean-burning fuels weighed in in support of the much stricter federal ozone and particulate standards that were adopted by the Clinton administration in 1997. By late 1998 two oil companies, three electric power companies, and more than a dozen other U.S. corporations publicly promoted steps to reduce global warming, and several announced quantitative goals for their operations to reduce greenhouse gases.[117]

Market pressures to reduce pollution may increase as the public gains access to more corporate information. In general, capital markets still ignore the value of environmental initiatives that go beyond government requirements, and the Securities and Exchange Commission so far requires only minimal disclosures about possible environmental liability. But investors do want to know if firms have plans to comply with new as well as existing requirements, and the government is considering increasing reporting requirements for company environmental information.[118]

Green-market niches sometimes bring in revenue, though evidence is mixed about whether and how much customers are willing to pay for products that are themselves environmentally friendly or are produced in an environmentally friendly manner. Interface Inc., the world's largest commercial carpet company, reportedly is invited to bid on big jobs for Gap Inc. and some other companies because of its environmental initiatives to reduce waste and recycle its products, measures the company says have also saved $25 million since 1995.[119] Some firms dealing in forest products in international markets have found that customers will pay a little more for wood certified by an international Forest Stewardship Council, especially in Europe, where eco-labeling is popular. The deregulation of electric utilities in many states may increase consumer options to choose "clean" electricity. When large corporations do choose to buy or sell "green," the effect of their decisions is multiplied by their influence over thousands of suppliers or customers.[120]

Particular business requirements—for natural resources, skilled labor, or improved efficiency, for example—produce diverse approaches to environmental issues. Some businesses benefit directly from pollution control and conservation. Tourism is a commonly cited example of a business that gains from protection of natural resources (though recreational use can also pose threats to wilderness, the seashore, and other protected areas). Tourism accounted for nearly 10 percent of U.S. jobs in 1995, and the industry wields increasing political power.[121] Companies that help make tourism the third largest business in Texas also helped persuade voters to amend the state constitution in 1995 to let ranchers keep their agricultural tax exemptions if they returned their land to its natural condition, a boon to hunters and bird-watchers.[122] Other businesses have a direct interest in pollution control or conservation. Firms that require large amounts of pure water—computer-chip manufacturers, food-processing companies, and breweries, for example—have an interest in keeping streams, rivers, and groundwater uncontaminated.

Particularly when high-tech industries are gaining in economic importance and skilled labor is in short supply, measures that improve an area's appeal as a place to live or work gain value to some employers. In the late 1990s the shortage of skilled workers slowed economic growth in southern states.[123] Some experts believe that the needs and availability of a skilled work force are replacing cost as the main factor in business relocation decisions, and there is evidence that skilled workers attach economic value to areas with less pollution.[124]

The most promising long-term trend, though, is the reduction in use of natural resources and the reuse of wastes by businesses that are searching for greater efficiency. While federal rules have emphasized end-of-the-pipe controls, substantial environmental gains have resulted from innovations like the radial tire, the replacement of copper wire by optical fibers, and the replacement of glass bottles by aluminum cans. Jesse H. Ausubel, director of the Program for the Human Environment at Rockefeller University, argues that further efforts by business to use wastes from one process as resources for another, to reduce the use of materials that pollute, and to produce energy from low-carbon sources can lead to major cuts in pollution. Ausubel points out that such trends have long been under way in industrial countries, highlighted by the declining use per unit of carbon-intensive fuels and the declining use of water. Federal efforts to encourage advances in such "industrial ecology" are coordinated by the Department of Energy through the Lawrence Livermore National Laboratory.[125]

Improving efficiency sometimes reduces pollution or furthers conservation directly. Installing a more efficient turbine in a coal-burning power plant, for example, both saves on fuel costs and directly decreases air pollution. So does improving the efficiency with which gasoline powers auto engines or finding ways to ensure that more pesticides reach pests. Uniform practices can also improve both efficiency and environmental protection. The OECD reports that multinational corporations "often adhere to a single (high) standard in *all* their operations world-wide" because designing to several standards is less efficient.[126] Some businesses also improve environmental practices to reduce uncertainty. The unpredictable course of new regulation can give businesses an incentive to adopt rigorous standards in advance of government action or to negotiate pockets of certainty. Since 1992, for example, many developers and other commercial enterprises have agreed to set aside endangered species' habitats on their property in return for government guarantees of "no surprises" in future regulations.[127]

On the other hand, businesses sometimes have particular reasons to resist further government action. Each additional unit of environmental protection becomes more costly as easy problems are solved. Increasing national and international competition adds pressure to reduce costs. Where competition is greater, it may become harder for business to pay for improved environmental protection by raising prices, meaning that shareholders will pay with reduced profits or employees will pay with forgone wage increases. And new government rules can mean mountains of paperwork, heavy demands on executives' time, and complex legal questions.

Business initiative may also be discouraged by the end-of-the-pipe focus of most regulation and by the paucity of common benchmarks for environmental management. Government technology standards can lock in practices or equipment that impair future efficiency. Firms have trouble tracking the real benefits of pollution control when they are indirect or occur at a later time and are not encouraged to do so by Wall Street. Many firms still view environmental protection as a social tax rather than an ordinary cost of doing business or an opportunity to gain efficiency. A recent broad analysis of environmental protection, including interviews with many business executives, found that most businesses still focus on the costs of environmental compliance rather than

87

on the business value of environmental performance, which is difficult to measure. Recent government and industry initiatives are attempting to address these problems.[128]

Leadership matters. A far-sighted chief executive officer can transform company environmental practices. Chief executives of Dupont, Monsanto, Xerox, Minnesota Mining and Manufacturing (3M), and some other large firms have made pollution control a high priority, have advanced environmental management, and have influenced suppliers. Dupont has announced that the company will eliminate pollution altogether, hoping to improve efficiency by cutting waste.[129]

As environmental policy matures, some of the toughest environmental issues arise among the smaller, more widely dispersed firms that account for most U.S. economic activity. While the few large and visible companies with national or international markets have increasing incentives to comply with environmental rules, smaller businesses may lack the motivation and resources to comply. Big companies tend to have more at stake in their public reputations than do smaller, less well-known firms. They may also have more reasons to plan ahead than small firms that do not expect to stay in business permanently. By and large, small businesses have lower capitalization and less stable profitability than large ones and are more likely to fold or suffer extreme economic consequences from new environmental costs. Congress acknowledged those problems when it passed the Small Business Regulatory Efficiency and Fairness Act in 1996, requiring agencies to consider the effects of their actions on smaller firms.[130]

In sum, the important business issues of the present and future do not jibe with the symbolism of the past. Images of large factories belching smoke and discharging chemicals into rivers that still animate debate in Washington should be revised to include the more intransigent pollution problems of dry cleaners, gas stations, farmers, and other medium-sized or small enterprises. Images of clear-cutting in national forests and overgrazing on federal rangeland should be updated to reflect the more complex problems of conservation on farms and in urban areas. A national policy framework constructed to address big business recalcitrance and conservation problems on public lands now struggles to contend with problems of a very different sort.

After 30 years, assumptions that provided the original underpinning for a national agenda need to be revised. A sense of national crisis has been replaced by an enduring public commitment, but without an accompanying resolve to change habits or pay the public bill. The idea that bold national policy required a system of strict legislated deadlines and uniform directions that tied the hands of bureaucrats has been replaced, for the time being, by congressional gridlock and administrative experimentation, amid signs that the real influence of federal rules may be diminishing. The idea that states engage in a "race to the bottom" has been replaced by a contentious contest to gain prosperity in a period of economic change, in which some environmental protection measures are seen as good investments and others are seen as liabilities. And the assumption that big business has unlimited capability but lacks the commitment to reduce pollution and improve conservation has been replaced by generally improved business practices in a time of declining business security, a wide variety of business responses to particular issues, and increasing public concern about small business and farms. Chapter 4 looks more closely at the new environmental politics emerging from these changes.

4

The New Environmental Politics

If you have traveled to the remote parts of the Deep South, I am sure you have seen the architecture of Tobacco Road—shacks built of whatever materials were available at the time, often by a series of owners. Maybe the roof is corrugated tin, but one wall is made from a billboard and the door step is a cinder block. No part matches any other part, and there are holes here and there. Still, it provides a measure of basic shelter, and there comes a point where it is easier to tack a new board over a gap that appears than to redesign the entire structure.[1]

NEARLY 30 YEARS AGO Congress constructed a national framework of policies to protect the environment. It rested on a foundation of ideas that reflected the political and economic currents of that time. The structure has now been expanded and remodeled by 14 Congresses and six Presidents. Its existence is remarkable. But it is also ungainly and complex, and everyone agrees that it suffers from structural cracks. The foundation has shifted as times have changed and the building needs to be adapted to new underpinnings and new uses.

Unheralded by public events or major legislation, that process of adaptation is well under way as the nation enters a new century. Debate about environmental policy often remains polarized in yesterday's ideological terms, studded with allusions to national crisis, bungling bureaucrats, state incompetence, and corporate evils. But, as a practical matter, the remodeling of the nation's approach to pollution control and conservation is visible everywhere. Across the country, immediate needs to resolve unfamiliar conflicts between economic and environmental priorities have forced opposing groups to hammer out compromises that address evolving problems and try out new public and private roles, even without congressional action to revise major laws.

Pragmatism is the unifying theme of this new environmental politics. The existing framework of environmental laws and regulations remains, but it is now accompanied by a latticework of informal accommodations to changing problems and changing times. In a subtext of political battles rarely visible to the public, federal and state agencies are attempting to sort administrative responsibilities on the basis of the apparent strengths of each level of government. Policy instruments are evolving to provide updated means of furthering common goals in diverse situations. Some national standards are quietly being customized to suit particular geographic or industry circumstances. Governments at all levels are making more use of financial incentives to reward or penalize private actions. And they are becoming more aggressive in enlisting information disclosure as a regulatory tool, aided by the revolution in computer and communication technology.

A Pragmatic Approach to Federal and State Responsibilities

The outdated idea that states engage in a "race to the bottom" to minimize environmental protection still has currency in a national debate that remains destructively polarized. Nonetheless, federal and state officials often are taking a pragmatic approach to dividing responsibilities, when current laws allow it. Nearly all environmental issues that draw national attention are, in fact, a mixture of local, state, and national (and sometimes international) concerns, so deciding *how* to share responsibility rather than *whether* to do so is the important issue. Consider these examples:

—With respect to assigning federal and state responsibilities, air pollution is by its nature a policy enigma. Carbon monoxide, one of the six common pollutants regulated by the federal government, usually drifts only a few blocks from its source. From that point of view, its control could be considered purely a local issue. But its source is primarily cars that move from state to state and are sold nationally and internationally by relatively few manufacturers. Pollutants that contribute to ozone, by contrast, are blown hundreds of miles from the power plants, factories, and cars that produce them. But they may change chemical character along the way, and their path and range cannot be ascertained with precision. Also, while it is true that most air pollution is concentrated in urban areas and remains local, more than a third of metropolitan areas cross state lines. To complicate matters further, gases thought to contribute to global warming confound conventional notions of where control authority should be placed because they affect temperatures everywhere on the earth's surface, no matter where they are emitted. Carbon dioxide, a colorless, odorless gas predominant in trapping heat in the atmosphere, is produced, literally, by everyone. Humans and animals exhale carbon dioxide, fossil fuels produce large quantities of it when burned, and trees and other plants absorb it in daylight.

—The disposal of hazardous waste is arguably a purely local issue. But waste at one facility often originates from many states, creating an expensive web of interstate litigation when something goes wrong. Interstate transportation of hazardous waste is itself a big business (separately regulated by the federal Department of Transportation). And inadequate disposal can, of course, pollute the air or interstate waterways or underground aquifers, the largest of which lies beneath eight states.[2]

—Drinking water safety also is arguably a local issue. If city or town residents choose to spend money on schools instead of a new filtration plant, why should the rest of us care? It may be true that the causes and effects of *chemical* contaminants are in many cases local. But it is hard to argue that *biological* contaminants are purely a local matter. Contagious disease is one of the nation's oldest and most basic public health concerns, heightened now by increased travel. And chemical contaminants from portions of watersheds located in neighboring states, from pesticides sold nationally, or from toxins blown into lakes and reservoirs from distant sources cannot be controlled effectively by any one jurisdiction. The degree to which drinking water is contaminated from out-of-state sources is a factual question that is worth asking but is hard to answer because sources are locally unique and because much remains to be learned about the migration of chemicals in both surface and groundwater.

Today's piecemeal assortment of federal and state laws and regulations will never be replaced by a rational structure. The character and strength of environmental programs within a state and among states inevitably vary. The interplay of federal and state (and regional) interests in framing particular laws and regulations is, of course, a political process, influenced by chance combinations of leadership, lobbying, economic trends, scientific evidence, pollution incidents, and many other forces. Over time, policymakers adapt to changing circumstances by means of ad hoc compromises, with varying degrees of success.

But, as a substitute for outdated "race to the bottom" reasoning and an illustration that more pragmatic approaches are possible, it may be useful to summarize some factors that may influence the strength of federal and state interests in addressing particular environmental problems, and to suggest how those interests interact to produce more aggressive or more limited policies. In practice, the issue is usually not *whether* each level of government will take action, but how aggressively each will intervene.[3]

From a pragmatic perspective, state decisions about how aggressively to address an environmental problem may be influenced by a variety of factors. They include whether the government is capable of action, whether benefits are concentrated within the state, whether benefits appear to be directly related to human health or to ecology and will be experienced by current residents, and whether action contributes to the government's vision of how to gain prosperity (see chapter 3).

Parallel considerations may influence the aggressiveness of action by the federal government. Congress and the president may have stronger incentives to overcome political obstacles if the federal government has reasonably clear legal and institutional capacity to tackle a problem, and if benefits will accrue within national boundaries but are not expected to be concentrated within a single state. Benefits may be spread among several states, may be too indirect or uncertain to prompt state action, or may go to residents of states that have not paid for improvements. The federal government also may be more inclined to take relatively aggressive action if proposed measures have strong links to improving human health or ecological concerns, at least in the long run, and if action contributes to the nation's current vision of how to achieve prosperity.

There are several reasons why the federal government may value distant and indirect environmental benefits more highly than do states. Traditionally, the federal government has had considerably more control over its competition with other countries through tariffs and other legislated conditions on trade than states have over their competition with other jurisdictions. It has more borrowing power and therefore historically has been less constrained by current-year resources. Better able to muster research and staffing resources, Congress and federal agencies also have tended to have better command of advancing scientific and technical knowledge about distant or complex health and ecological consequences than do many state legislators and agencies. Also, the federal government generally has been lobbied more vigorously by environmental and scientific groups that promote a longer time horizon. Voters' assumptions about the mobility of the American population may also favor a longer federal perspective. Voters inclined to invest in a better environment for their children or grandchildren may not be willing to invest in the notion that their progeny will stay in the same state but may assume that they will at least reside somewhere in the United States.[4]

Viewing environmental policies pragmatically, as reflecting different combinations of federal and state interests, may be helpful in two ways. It may underscore an obvious but often neglected point: federal and state interests can reinforce each other to produce aggressive action by both levels of government, can reinforce each other to produce relatively limited action, or can produce aggressive action by one level of government and weak action by the other.

Such a pragmatic assessment might also provide federal policymakers with some clues about when states are likely to carry out federal priorities with diligence. For example, in cases where both national and state interests in action are strong, effective measures might consist of clear federal goals and maximum state flexibility in carrying them out, on the grounds that states can be counted on to monitor and enforce federal policies where their own interests are at stake. If national interests in action are strong and state interests are weak, progress might be better served by clear federal goals and limited state flexibility in carrying them out. If national interests are weak but state interests are strong, effective policy might involve limited federal actions such as guidelines, research and development funding, technical support, and public information accompanied by aggressive state action. Finally, if both national and state interests are weak, issues might involve little or no action by either level of government.

Pragmatism in environmental politics also invites new interest in regional approaches to pollution control and conservation. The potential benefits of regional cooperation receive too little attention in a political system that emphasizes national and state authority. Many water quality issues are by nature regional, shaped by the chance contours of watersheds, rivers, lakes, and underground aquifers. Many air pollution problems also are regional, influenced by geography and weather patterns. Some of today's most serious environmental challenges, including problems caused by polluted runoff and threats to entire ecosystems, are more sensitive to regional differences than those of the past.

In an effort to cope with dispersed sources of pollution and conservation issues on private land and to make use of the growing scientific understanding, both Congress and the executive branch have begun to take a greater interest in regional approaches. The Department of the Interior is dividing the country into ecologically defined regions and is prioritizing areas for conservation.[5] In 1997 the EPA initiated a watershed approach to controlling water pollution that was reminiscent of regional approaches in the 1960s, before federal laws nationalized the issue. The EPA's watershed approach calls on federal, state, local, and industrial representatives to identify pollution from all sources and agree on controls. As elaborated in a Clean Water Action Plan in 1997, it encourages state-led assessments to set priorities, anticipates expanded federal criteria for stormwater and nutrient runoff in 1999 and 2000, commits the Department of Agriculture to fostering 2 million miles of conservation buffers by 2002, and promises increased federal funds to aid the effort.[6] Starting in 1995, 37 states and the District of Columbia worked for two years under the EPA's auspices to find ways to reduce the drift of ozone-causing pollutants over the eastern United States. Despite bitter conflict between upwind midwestern states and downwind eastern states, a majority of representatives agreed in principle that smog-related pollution from power plants should be cut by 85 percent and such pollution from factories by more than two-thirds. In 1998 the EPA used the regional group's work as a basis for strict new requirements for 22 states and the District of Columbia to reduce nitrogen oxides—by 51 percent in West Virginia, for example, and by 36 percent in Ohio and Indiana—despite protests by midwestern governors and power plant executives.[7] New initiatives build on some long-standing regional approaches, such as interstate efforts to clean up the Great Lakes, the Delaware River Basin, and the Chesapeake Bay.

Sometimes, regional cooperation can spread the costs of addressing a complex problem or concentrate pressure for action among affected states. But such cooperation also has its limits. For one thing, it is unlikely to alter fundamental state interests (see chapter 3). For another, regional approaches are politically awkward. They superimpose on a constitutional system of federal and state government a plan of action that fits neither. Strong common interests in regional decisions may overcome political obstacles on occasion. But if such decisions are a good match environmentally, they will always be something of a mismatch politically.

Even when the practical forces at work in resolving issues of federal and state authority are fairly well understood, results may not be predictable. But introducing a template for assessing the interests of different levels of government in environmental protection case by case does offer an alternative to the outdated notion that irresistible economic forces cause states to minimize all such efforts in a "race to the bottom."

Customizing Policy

As pragmatic approaches gain ground, the relatively simple structure of uniform standards and strict deadlines mandated by Congress in the early 1970s is being overlaid by a much more complex pattern that includes national rules, as well as rules and agreements customized by state, locality, or region; by industry; or by source. This piecemeal evolution, rather than the dim prospect of wholesale reform, is one place to search for clues to the kinds of policy that may be effective in the future. It also calls for vigilance about the fairness and equality of requirements and about the effectiveness of public participation. Some of the changes that have occurred are a result of Congress's efforts over the years to accommodate unavoidable geographic or industry differences. Others have been initiated by federal and state officials as administrative adjustments, demonstration projects, or exceptions to usual rules to resolve urgent conflicts between economic development and environmental protection or to provide a viable approach when formal rules seemed inadequate and Congress was unable or unwilling to act.

Setting clear national goals and giving states and businesses as much flexibility as possible in how they are carried out is one way to mediate between broad priorities and specific actions. Several thoughtful critiques of national policy have urged wider use of such performance standards.[8] Where feasible, such standards have special appeal for integrating national and local concerns in dispersed, site-specific environmental problems such as water pollution from farms and city streets. A performance standard sets *the level* of pollutant or ecological indicator that is required without dictating *the means* by which that level should be reached. Since the early 1970s, some performance standards have been included in national policy. These include ambient air quality standards and drinking water standards that limit the allowable concentration of contaminants in air and water. Design standards, by contrast, require a particular kind of equipment or process to reduce pollution or improve conservation. A 1977 legislative compromise that required new coal-burning power plants to install "scrubbers" in smokestacks to remove pollutants was one such design standard.[9]

In practice, however, the distinction between performance and design standards is not simple. Federal drinking water standards set performance standards for contaminants. But federal officials in fact promote particular means of achieving those standards. In 1998, for example, the EPA sued the state of Massachusetts to force the construction of a new $180 million filtration system. Federal officials characterized as inadequate a state plan relying on ozone treatment and land use controls to produce safe drinking water.[10] Many federal air and water pollution requirements for businesses are, in effect, design standards, requiring or encouraging the installation of particular equipment.

Uncertainty, itself, can generate controversy. Questions about what to measure or how to measure it can lead to political battles. In 1995, for example, federal and state officials fought over whether to change congressional directives that the nation's 200,000 public drinking water systems test for 83 contaminants, with 25 more added every three years, even if they had never been found in local drinking water. By the mid-1990s many states were reporting that their water authorities lacked the resources to carry out critical tasks to ensure that drinking water was safe because federal rules forced them to spend much of their limited resources on testing for unlikely contaminants and on other tasks less important to that goal. The upshot was that Congress amended the law in mid-1996 to allow the EPA to weigh costs and benefits when setting future standards and to give small communities somewhat more flexibility in testing and monitoring, though the law also added new requirements that the Congressional Budget Office said might triple costs of carrying out national rules.[11]

Performance standards have the obvious advantage of allowing firms or state governments to decide the most effective way to meet national goals, whereas design standards tend to lock in technologies that may become outdated in future years. But performance standards work well

97

only where pollution or ecological processes are well understood and where accurate measurement is feasible, making monitoring and enforcement of standards realistic.

A related, broadly supported policy shift goes further, replacing the idea of uniform national rules with different standards for different communities or industries, at least in the short term, justified by their variable circumstances. Over time, for example, Congress has acknowledged the inevitable influence of variations in geography, history, and economic circumstances on air pollution in urban areas. Beginning in 1977, Congress replaced uniform air pollution standards with different phased reductions for heavily polluted metropolitan areas (called "nonattainment areas"), giving them more time for cleanup in return for more vigorous remedial action. In 1990 Congress turned that idea into a five-tier ranking of urban areas (severe, extreme, serious, moderate, and marginal pollution), with different annual goals for each.[12]

Several obstacles block a more decisive move toward performance standards, which are supported in principle by both Democrats and Republicans. One is a lack of agreed-upon pollution measures and ecological indicators, and the fact that many environmental processes are still imperfectly understood, even after a generation of general progress.

A second obstacle is political in nature. Institutional pressures within federal agencies favor centralized control, and legitimate concerns about accountability favor continued federal rules that specify the means by which national goals are achieved. For 30 years, for example, Congress and agency officials have specified not only limits on auto pollution, but also how those limits should be achieved. In the 1990s the EPA required some heavily polluted areas to adopt centralized inspection and maintenance programs for cars already on the road. Reacting to drivers' objections to the expense and inconvenience of central inspection, however, many states favored checkups at neighborhood gas stations or other approaches. The result has been a standoff, culminating in the outright refusal by several states to carry out advances in the program.[13]

In the past decade, environmental protection has moved more often to the bargaining table, raising hopes of a less adversarial process tailored to suit particular circumstances. In the 1990s the national effort to protect endangered species on private property became, in part, a giant checkerboard of agreements involving property owners, government agencies, and private groups, placing ecological constraints on the use of private land. Beginning in 1992, the Department of the Interior took on the task of negotiating nearly 500 habitat conservation plans with landowners, covering more than seven million acres by the end of 1998, to protect endangered species on private property. It did so by securing legally binding promises to protect habitats in return for government commitments of "no surprises" in new regulation, even if circumstances changed. A 1998 survey by the Land Trust Alliance, an association representing conservation groups, estimated that land protected by private national groups like the Nature Conservancy had doubled in the past 10 years to 13 million acres. Local land trusts protect another 4.7 million acres, mainly as a result of negotiated conservation easements with landowners.[14]

Likewise, having constructed an intricate framework of rules and procedures for states and industries over a period of 25 years, EPA officials in the 1990s encouraged them to join in negotiating shortcuts through that complicated maze of requirements. Beginning in 1995 EPA officials negotiated with about half the states individually tailored plans for pollution control called National Environmental Performance Partnership Agreements, exchanging somewhat more flexibility in meeting national requirements and a promise of less detailed federal oversight for new commitments to improved performance, improved state management, and public participation.[15]

Federal officials have also experimented with replacing uniform procedures for pollution control with agreements customized by industry, or by manufacturing plant. EPA officials agreed to ignore some review and reporting requirements for a handful of big businesses that agreed to superior pollution control performance and experimented with industry-wide goals to replace separate regulatory structures for air, water, and land pollution.[16]

Experiments with negotiated protection appear to be driven mainly by growing political pressures to increase the speed, certainty, and efficiency of rules. By the mid-1990s government officials and business leaders shared a growing frustration with the time and expense involved in making, complying with, and enforcing increasingly complex national and state standards and procedures. When leading Republicans in the 104th Congress tried to translate that frustration into a partisan effort to weaken federal requirements, they intensified the need for a response by the Democratic administration. That response had to feature steps that could be carried out without seeking large appropriations or new legislation from a Republican-controlled Congress. Such experiments also benefited from changes in public attitudes that have made it politically possible for federal officials to try out more cooperative approaches with states and businesses (see chapter 3).

It is too early to know whether these tentative steps are part of a transition to new approaches to pollution control and conservation or are simply a temporary means of relieving political pressures. They are countered by legitimate fears that they may in fact weaken environmental laws, and by institutional pressures to keep central control in Washington. At times, EPA officials have appeared to backtrack from commitments to state flexibility, agreeing to only minor changes.[17] And the popularity of negotiation has not stopped federal officials from debating the minutiae of pollution control procedures with state representatives.[18]

Whatever the future of such agreements, a few observations may help place them in perspective. First, negotiated protection is not new. Conflicts between economic development and environmental protection have always been resolved mainly by negotiation. The long process of informal rule making, orchestrated by the federal Administrative Procedures Act, often amounts to highly structured bargaining. Litigation also is a forum for negotiation and usually ends in settlement. More than 90 percent of environmental enforcement actions are resolved before trial, by consent decrees, the use of administrative enforcement procedures, or other means.[19] What has been different about the increasing popularity of negotiated protection in the 1990s is its ad hoc character. Agreements with states or industry have often avoided established rules, and negotiators have invented new procedures along the way.

Second, the advantages of negotiation in some circumstances and the frustration with the complexities of federal standards should not obscure the fact that the more adversarial system of rule making has many strengths. It was chosen in the 1970s because, at its best, it provided clear, lasting decisions through an open process structured to provide an opportunity for broad participation. In practice, that opportunity has often been gained by sacrificing speed. Complex rule making may take as long as four or five years, and the EPA has usually failed to meet statutory deadlines. Also, the certainty and accountability of some major standards have been undermined by years of lawsuits resulting in complex judge-made law. But if both process and results have fallen short of expectations, such rule making also has fostered much progress. For the foreseeable future, it remains the dominant means by which public sentiment for improved environmental protection is translated into action.

Third, negotiated agreements are often useful, but they are no panacea. They can impose a new layer of requirements that amounts to government-by-memorandum-of-understanding, can be removed from public view, are open to charges of inequity, and can be too voluminous and complicated to enforce effectively. Negotiation does not necessarily improve the speed and certainty of decisions, either. One study of "reg neg"—that is, modified rule making that emphasizes early negotiation—found it was little used and did not produce quicker, less litigated results. Ad hoc negotiation also can create new rounds of litigation, especially where procedural issues are novel.

Policymaking by negotiation also can awaken old fears. Voters in the 1990s sometimes found it hard to distinguish between federal and state attempts to improve regulatory effectiveness and attempts to grant business new favoritism. Environmental groups understandably have been skeptical of approaches that give individual companies special treatment. Also, final terms can be costly to taxpayers. In the late 1990s several giant land negotiations intended to promote conservation hinged on substantial federal commitments to buy property, commitments that Congress has been reluctant to expand.[20]

In fact, rule making and ad hoc negotiation can be complementary. Rule making tends to work best when there are many interested parties, uniform results are desirable, and winners and losers are inevitable. Negotiations tend to work best when there are relatively few interested parties, uniform results are not needed, and all participants stand to gain by an agreement. The two regulatory techniques also are frequently interrelated. For example, a fear of rule making sometimes encourages businesses to engage in informal negotiations.

These moves toward performance standards, negotiated agreements, and regional actions are signs that the political system is adapting to changing problems and changing times, even in the absence of congressional action. They are worth watching for two reasons. First, they may provide clues about more effective ways to reduce pollution and improve conservation. Second, they test the political viability of alternative approaches.

Using Financial Incentives to Influence Choices

In January 1998 President Clinton proposed a striking new environmental initiative. The national goal was to cut carbon dioxide and other greenhouse gas emissions to reduce the risk of global warming in order to meet specific international targets, a commitment made by the United States in Kyoto, Japan, a month earlier. Industry would have to make quick reductions in its use of coal, oil, and other fossil fuels. Drivers and homeowners would have to cut down substantially on their use of energy. Such sudden changes might seem to call for strict national standards with harsh penalties, the kind of approach Congress employed to fight pollution in the 1970s. But instead the president proposed an array of tax credits and government grants to reward businesses and the public for changing their ways quickly. These measures included a 10 percent tax credit for industries that invested in more efficient energy use, a 20 percent tax credit for building owners who installed advanced heating or air-conditioning systems, and a tax credit of at least $3,000 for purchasers of cars that get double or triple the usual gas mileage. The cost to the national treasury in lost revenues and direct spending was estimated at $6.3 billion over five years.[21]

Taxes, subsidies, emissions trading, and other financial incentives that directly increase the cost or rewards of private choices are moving into the mainstream of environmental policy. They are a supplement to, not a replacement for, government rules. Rules and financial incentives, like rules and negotiated settlements, can work together. Sometimes they are used in combination, with rules defining boundaries within which incentives are applied. Sometimes they are used in tandem. Rules tend to work best when uniformity is feasible and when predictable results are needed. Financial incentives work best when sources are many, uniformity is not needed, and variable results are appropriate. (Both standards and incentives work well only if the causes and effects of pollution are known, and policies are enforceable.)[22] In national, state, and local actions, such incentives have gained an important role in improving environmental protection.

There is nothing new about government use of financial incentives to influence resource use and commercial plans. Federal law has long included a labyrinth of subsidies, taxes, tax credits, grants, loans, and penalties that influence the choices of businesses, farmers, and individuals regarding economic gain and environmental protection. To cite just a few examples, federal subsidies have sometimes encouraged timber-cutting and grazing on public land, agricultural subsidies have long encouraged farming of marginal lands, and 60 years of federal mortgage assistance has encouraged the construction of new homes.

Some traditional incentives are now being refashioned to give more weight to environmental priorities, while continuing to support other national goals as well. Federal subsidies to farmers, for example, have become increasingly focused on environmental objectives. Since 1986 federal payments have encouraged farmers to convert 36 million acres from agricultural production to erosion control and wetland protection by growing grass and trees. The same Conservation Reserve Program is used to subsidize farmers who grow filter strips of vegetation along waterways, so as to reduce polluted runoff from fields.[23] In 1989 Congress imposed a tax on the production of chlorofluorocarbons so as to phase out their use, in accordance with an international agreement. On the other hand, Congress has so far declined to follow the example of most European countries that limit energy use and pollution by placing taxes on gasoline that are greater than its market price. In the United States, gasoline is currently among the cheapest liquids sold; it is less expensive per gallon than soft drinks or brand-name bottled water, for example.[24]

An authentically American invention of the past 20 years has been the mechanism of emissions trading as a means of encouraging businesses to reduce pollution in a cost-effective manner. Emissions within a geographic area are capped in legislation. Companies are granted permits for units of pollution (by grandfathering existing pollution or by auction) that can then be bought and sold among polluters. Trading has been used successfully in some of the relatively rare situations where such markets can be created. The 1990 Clean Air Act set up an innovative trading system for sulfur dioxide emissions to reduce acid rain. Preliminary evidence indicated that it might save the country $4 billion to $5 billion a year, in comparison with a technology-based approach. In the 1980s lead rights were traded as part of a program to phase out leaded gasoline. An innovative trading program in the Los Angeles area caps emissions for nearly 400 industrial polluters and decreases allowable pollution each year.[25] The multilateral Kyoto agreement in 1997 concerning global warming included an option of international trading in greenhouse gas emissions. In 1998 the EPA endorsed state trading systems for nitrogen oxides as one means of achieving a 28 percent reduction in those smog-related emissions by 2007.[26]

State governments also use financial incentives to further environmental objectives. A number of states grant tax credits to encourage the cleanup and redevelopment of abandoned industrial sites. Illinois, for one, initiated a 25 percent tax credit on cleanup costs of up to $700,000 in 1997.[27] New York gives tax credits to trucking firms for converting their fleets to natural gas, refunding 60 percent of conversion costs.[28] More than 2,700 localities charge residents by quantity for throwing away garbage, a number that appears to be increasing.[29] In Texas, a recent constitutional amendment lets ranchers keep their agricultural tax exemptions when they use land to encourage wildlife instead of to graze cattle.[30] A number of states have set up wetlands banking systems, allowing landowners to buy credits from groups who protect wetlands throughout the state, programs encouraged by the federal government.[31]

One danger, of course, is that new incentives can be layered on top of existing government subsidies that work at cross purposes, instead of being integrated into a coordinated policy that balances competing interests. Long-standing federal tax policy encourages employers to offer free parking. But President Clinton's global warming program calls for the subsidization of employee transit fares. Long-standing agricultural policy encourages farmers to enrich soil in order to boost productivity. But recently, several states offered tax incentives to cut down on fertilizer use in order to reduce nitrogen runoff, a leading cause of water pollution. Timber and mining industries together receive $1 billion in annual tax benefits to encourage production. But federal regulators have been cutting back production to encourage conservation. Assessing such federal subsidies would be a complicated but constructive first step, as suggested by the President's Council on Sustainable Development and by Enterprise for the Environment, a broadly based panel of policy-makers, environmentalists, and business representatives.[32]

Taken together, a variety of financial incentives hold considerable promise, but their effectiveness depends on a clear understanding of their strengths and weaknesses. The EPA's Science Advisory Board summarized some of the factors influencing the effectiveness of different types of incentives in a 1999 report analyzing environmental risks. The board noted that incentives that influence *prices* work best where habits are influenced by changes in price, variations in the results are acceptable, and performance can be tracked. Most financial incentives now in use are price based, including subsidies to farmers, tax credits to firms to encourage energy efficiency, trash fees based on volume, and soda-can-refund programs.

By contrast, tradable permit schemes that influence private choices by placing a ceiling on the *quantity* of pollution allowed work best in those relatively rare situations where there is a regional group of sources, the costs of improving practices are expected to be high and variable, sources are identifiable, property rights can be defined in a way that allows a market to function, and progress can be monitored.[33]

But incentives, like regulations, can miss the mark. They can lock in a sensible approach to yesterday's problems, can be difficult to set at levels that influence decisions in intended ways, can focus attention on the wrong target, or can shift pollutants around in ways that regulators have not anticipated. Tax incentives to encourage the purchase of pollution-control equipment, for example, can discourage efforts to redesign production so as to reduce pollution's incidence, and taxes on waste disposal can lead to midnight dumping. Mistakes are hard to correct. Once incentives are adopted and acquire beneficiaries, they are hard to change.

Sometimes, too, the flexibility that incentives offer is not desirable. When given choices about how much pollution to reduce, firms in different areas will make different decisions, undermining the notion that every community deserves a minimal level of clean air and water. For that reason, incentives are also counterproductive when human health is immediately threatened. And, like rules, incentives create winners and losers among firms; in some forms, they can systematically favor more profitable enterprises that benefit most from tax credits, for example.

Employing Information as Regulation

Now that environmental values are embedded in national and state politics, regulators use information requirements themselves to create incentives for businesses, individuals, or government agencies to limit pollution and further conservation. Regulators hope that businesses will respond by improving their practices, in order to improve efficiency, gain customers (or votes), or avoid bad publicity. Enhanced by computers and communication technology, information collection and disclosure are, in effect, a form of regulation. As such, they supplement, but do not replace, national and state rules, and, together with the use of financial incentives, signal increased government interest in employing market mechanisms to further environmental objectives.[34]

The idea of information as a kind of regulation is counterintuitive. People are accustomed to thinking of data as a technical tool supporting government action, not as a regulatory instrument. But at a time when national authority is challenged by changing political and economic forces, the government's unique ability to command the collection and use of information remains an enduring strength. National and state efforts to employ information as a regulatory tool have grown in recent years, emerging from searches for practical approaches to contentious environmental issues. Consider the following examples.

The 1996 amendments to the Safe Drinking Water Act, passed by the 104th Congress after two years of acrimonious debate, require 56,000 local water systems to alert customers to any violation of national standards with serious health effects; the systems must also notify customers once a year about bacteria and chemicals in tap water. Such notifications are intended to decrease contaminants by influencing the actions of local officials, making them more vigilant.[35] Using "Surf Your Watershed," an EPA Internet site (www.epa.gov/surf), anyone who enters a zip code can now get specific information about pollution sources, water quality, and drinking water sources. Also, a number of states require electric utilities to report environmental information.[36] Those provisions follow the example of the Toxics Release Inventory (TRI), a 1986 federal requirement that certain companies report on their discharges of toxic substances. The TRI is credited with contributing to a 43 percent decline in industrial toxic emissions from 1987 to 1997. Private groups now use the Internet to post and interpret TRI data by neighborhood.[37]

A number of states here and in Europe use spy satellite images purchased from countries such as India and Russia to encourage companies to keep their environmental commitments, a practice also common in Europe. Georgia's satellite monitoring of timber cutting is supported by some forest products firms as a means of allaying public suspicions about overharvesting, while Arizona compares satellite images of growing crops with irrigation permits in order to spot illegal water use.[38] If legal and political barriers to commercial high-resolution imaging from space are removed, its use for environmental purposes may spread.

In 1998 Illinois, Michigan, and Wisconsin negotiated commitments from Ford, General Motors, and Chrysler to provide over the Internet and on paper not only information about toxic releases, but also environmental inspection and compliance records for five factories. Federal officials are also working with companies to agree on release of factory information about other pollution and about violations of laws.

Both government and private groups are attempting to influence corporate actions by calling attention to firms that are leaders in environmental initiatives. The EPA rewards voluntary business efforts to control pollution with good publicity. Because the public is now attuned to the importance of environmental progress, business initiatives are considered news. When the Council on Economic Priorities, a New York nonprofit group, ranked Sun, Exxon Corporation, and Chevron Corporation as best among oil refiners in toxic emissions and waste, hoping to influence drivers at the pump, the *Wall Street Journal* noted the fact in a front-page story headlined "Sun Co. Shines in a Report Card on Oil Companies' Environmental Performance."[39]

Government agencies are also experimenting with the use of disclosure, where company data can be independently verified, to give companies additional incentives to meet federal and state goals. In an effort to control pollution from dispersed sources, some states have set ground rules that would enable dry cleaners, photo processors, and other small businesses to self-certify pollution levels.[40] Most national and state environmental laws already rely heavily on self-monitoring. But self-monitoring combined with verification and public disclosure may be a promising combination, especially when combined with computer and communication technology.[41]

Businesses themselves sometimes find it to their advantage to develop information that creates incentives to improve environmental protection. In January 1999, for example, power companies ran ads explaining efforts by the Electric Power Research Institute to come up with a formula measuring a plot of land's environmental importance. Such a formula, the ads said, would encourage utilities to trade ecologically important land to conservation banks and free them to develop property with little such value.[42]

A broader national and international trend among companies is the increasing use of environmental audits. In 1993 the European Union set ground rules for voluntary environmental auditing. It set forth management criteria, called for public disclosure of performance records and a commitment to continuous improvement, and offered companies technical assistance in preparing their audits. Under these rules, audits are to be verified by independent evaluations. The International Chamber of Commerce has also endorsed the idea of environmental audits, with the results reported to shareholders, employees, and the public.[43] Audits are most likely to be effective where companies have something to learn from the process and are likely to change their practices in response to the results.

In the United States, however, the potential use of company audits is entangled with liability issues, slowing progress toward agreed-upon standards. Because exposure to government enforcement actions may discourage companies from disclosing information about their pollution problems, at least seven states have enacted laws to keep audit information from being used in liability proceedings. Federal officials, who support the idea of environmental audits, have fought such state laws as obstacles to effective enforcement.

Public disclosures indicating that goods are being made or brought to market in an environmentally sound manner or that they themselves help reduce pollution can also create incentives for businesses to change their practices if customers respond to such information. Dolphin-safe labels on canned tuna fish, for example, changed fishing practices around the world because they influenced buying habits.[44] Growing concern about global warming and the international agreement signed in Kyoto, Japan, in 1997 produced new interest in "eco-labels." A month after the agreement was signed, electronics manufacturers announced that they would use a government-approved label to tell consumers that they had made a simple circuitry change to cut by more than half the electricity used by TVs and VCRs when they are turned off, thereby saving billions of kilowatt-hours of electricity each year.[45] To tackle dispersed sources of pollution that contribute to indoor air pollution, a recent article in *Scientific American* suggested that manufacturers include information about pollutants on product labels.[46]

Services, too, are promoted using eco-labeling. Utility deregulation creates opportunities for power companies to seek competitive advantage by promoting their environmental records. In California, some customers currently pay about 20 percent more for power that is accompanied by environmental assurances, though the number of customers making *any* choice is so far too small to draw conclusions.[47]

But will customers pay higher prices for environmentally friendly products and services? The answer may depend on such factors as the character of demand for the product, the perceived urgency of the environmental threat, and the nature of the label. Evidence from polls indicates that customers may be willing to pay a small premium for some environmentally friendly products, though they consider price, quality, and convenience more important. As environmentally conscious customers themselves, federal and state governments could exercise enormous direct influence on many markets. The federal government buys more than $200 billion in goods and services each year. A 1993 Executive Order directed the EPA to develop purchasing guidelines for federal agencies to minimize environmental harm.[48]

In an increasingly international economy, if major companies do decide to use eco-friendly products, they can create ripples of eco-friendliness among suppliers. A number of American forest products firms and at least four states have adopted timber-cutting practices that protect wildlife and as a result have won certification from the international Forest Stewardship Council. This certification has improved their ability to sell to major retailers like Smith and Hawken and Britain's Sainsbury supermarket chain, both of which purchase wood certified by the Council.[49]

The EPA Science Advisory Board has suggested that information works best as a regulatory tool where there are gaps in public knowledge that relate to increased risks, where information to fill those gaps can be easily collected and communicated, where availability of information will lead to change in practices that are causing the risk, and where variation in practices is acceptable.[50]

It would be a mistake, though, to view such use of information as an easy step. If eco-labeling is successful, it influences competitive advantage, so terms are hotly debated. A battle to define "organic" foods by federal regulation, for example, has been raging for eight years. In a 1998 skirmish, the Department of Agriculture proposed using the term to describe crops that were genetically engineered, irradiated, or grown in sewage sludge, and to describe meat from animals raised in confined conditions. This suggestion was met with the charge that such rules would give market advantage to corporate agriculture at the expense of small farmers, who define the term more narrowly.[51]

The potential of information to influence corporate or individual actions is as yet unclear. A great deal depends on the willingness of political leaders to wade into such treacherous political waters in order to set some ground rules for public disclosure that will ensure fairness and guarantee that data are always accompanied by neutral interpretation. Much also depends on the ability of public officials to frame requirements in a way that minimizes unintended consequences. Requiring public disclosure of hazardous discharges into air or water or on land, for example, can result in increased risks from wastes disposed of underground.

To conclude, the American political system is demonstrating its strength in adaptation. Although debate in Washington remains polarized and national legislation has been rare, new conflicts between economic and environmental interests have been resolved. Innovative approaches are being invented that may provide clues about the direction of future national policy. A general distrust of state and local government has been replaced with a more pragmatic approach to state interests and federal strengths. Uniform standards have been supplemented or modified in practice by a move to customize national requirements in order to meet the unique needs of different states, metropolitan areas, industries, or individual businesses. And a framework that relies mostly on rule making has begun to make greater use of financial incentives and the power of information to influence state, business, and individual choices.

Epilogue:
Turning Point or Stalemate?

TACKLING TODAY'S MOST SERIOUS environmental problems brings to center stage political battles that have raged at the perimeter of national policy for 30 years. Legislators trying to lessen water pollution caused by storm runoff or reduce threats to ecosystems clash with property owners trying to protect established uses of their land. Regulators trying to reduce auto pollution from aging cars and trucks vie with vehicle owners who believe that how much they drive and how they maintain their cars is their own business, not the government's. Public officials responding to voters' demands that neighborhood businesses should clean up toxic pollution battle with owners of gas stations or dry cleaning establishments.

In the past, successful remedies for concentrated pollution problems have relied mainly on *changes in technology,* often by few big businesses. By contrast, remedies for dispersed pollution problems and for widespread threats to watersheds or wildlife habitat rely mainly on *changes in behavior,* often by thousands of ordinary people. Problems arise in places where most people live and work, involve dangers that are often indirect, invisible, or little understood, and threaten long-accepted business practices, farming methods, and personal habits.

Technological changes, encouraged by government rules, have produced many lasting successes. The fact that people who live in Los Angeles can sometimes see the surrounding mountains, that major metropolitan areas experience fewer smog alerts, that lead is less of a hazard as an air pollutant, and that raw sewage no longer is dumped into lakes and rivers are improvements that would not have taken place without sustained public support for government action. There is no doubt that improving environmental protection has become a permanent national value.

Do new kinds of conflict signal a turning point in the nation's environmental policies, or a stalemate? Political and economic changes have certainly defined some limits to federal and state efforts, at least for the time being. Experience suggests that the public is reluctant to pay for all the improved environmental protection it says it wants or to change habits in ways that are personally inconvenient. Political and economic developments may be diminishing the practical power of the federal government to mandate environmental protection. Meanwhile, state budgets are being squeezed as a result of taxpayer revolts and declining federal support. The funding squeeze is most severe in poor states, where revenue-raising capabilities are limited, reliance on federal funds has been greatest, and environmental programs tend to be weakest. Competing in a fast-changing world economy, big businesses have new reasons to cut costs. And much future environmental progress depends on smaller businesses and farmers who have even less economic security.

Nonetheless, the need to resolve novel conflicts between economic and environmental interests in the context of new limitations has given rise to a new pragmatism in environmental politics. In the absence of congressional action to revise basic environmental laws, workable compromises to resolve emerging problems have been jury-rigged around and within the existing labyrinth of rules. Economic incentives are employed more frequently to further federal and state objectives. In Washington, debate often remains polarized. But around the country, the nation is in the midst of a rich, experimental time in environmental policy.

The single greatest obstacles to the continuation of such constructive adaptation are political and institutional roadblocks to expanding and sharing knowledge about the interaction between complex biological and chemical processes that support life and productive human activities that sometimes alter them. The case for emphasizing such efforts is particularly compelling now. Advances in computer and communication technology can enhance the power of public disclosure of information as a regulatory tool, and a generation of remarkable environmental research has created an explosion of knowledge about the earth. And in the 1990s, the federal government stepped up efforts to improve specific categories of information for policymaking, monitoring, and enforcement. In practice, however, the potential created by research and technology has often been countered by political pressures and institutional arrangements that tend to minimize and fragment usable knowledge. "Data to illustrate long-term, national emissions trends, concentrations in the environment, and human and environmental health effects are lacking for most . . . pollutants and media," one detailed study in 1998 concluded.[1]

Pragmatism cannot gain ground without reliable information to support decisions and to help create new incentives for government, businesses, and individuals to improve their practices. Yesterday's ideological choices were based on strongly held beliefs. Today's pragmatic choices stand or fall on the accuracy with which the sources of problems are assessed and the consequences of alternative actions are predicted. When reliable answers to pivotal questions are missing, debate tends to revert to a battle of preconceived notions. As policymakers tackle increasingly complex pollution and conservation issues, the information gap widens.

Although remarkable research has taken place and specific studies proliferate, many of the most basic questions about the nation's environment remain unanswered.[2] To cite a few examples, no one knows whether water quality is getting better or worse in most lakes, rivers, and estuaries. The migration and human health effects of hazardous air pollutants—those that may have serious health effects in small quantities—are still poorly understood. Although many new chemicals are being carefully screened, those that have long been in everyday use are only now beginning to be systematically checked for toxic effects.

Likewise, gaps in information threaten to overwhelm policies designed to address ecological concerns and reduce the prospects for reconciling conservation needs with development pressures. Despite continuing efforts to agree on indicators, the nation still has no accepted definition of ecosystem health. Precise vegetation maps have yet to be constructed for most parts of the United States, and reliable range maps are lacking for most animals. In 1997 the results of the first broad study of habitat conservation plans, the government's main vehicle for protecting endangered species on private land, indicated that crucial information about changes in populations and their habitats was missing for most species.[3]

For practical reasons, Congress and the executive branch have directed information-gathering efforts mainly toward immediate questions related to federal rules. Federal pollution control standards, for example, focus on emissions of a limited number of pollutants and concentrate on new sources. Most emissions information is collected by the sources themselves (usually businesses or local government agencies) and may be initially compiled in state reports, with no reliable basis of comparison from state to state. The fact that government methods of collecting information have changed over time and that they were hard hit by budget cuts in the 1980s and again in the early 1990s also has hindered progress. A National Research Council study in 1993 found that little biological information was used in government conservation decisions because the available information was not usually the sort that was helpful in decisions and because results were poorly communicated. Ironically, the result can be a wealth of specific information, but large gaps in basic understanding.[4]

Information gaps sometimes have resulted from a politics of willful ignorance. The collection and analysis of environmental information has been blocked by political interests that stand to lose (or fear they will lose) from the growth of knowledge. Information can be objectively presented, but it is seldom politically neutral. If information is power, that means it also poses a threat. In the 1990s, for example, environmental groups and government agencies interested in gaining information about biological resources and landowners interested in protecting privacy were an explosive political combination. In early 1993 a concerted federal effort to improve the quality of the nation's knowledge about the plants and animals on which human life depends by building a well-endowed National Biological Service in the Department of the Interior was shelved after a fierce campaign by property rights groups that feared such surveys would reveal endangered species on private land and restrict landowner plans. The service was given a more obscure name (Biological Resources Division) was stripped of most of its funding, and was placed under the auspices of the U.S. Geological Survey.[5] As a result of such political decisions, the United States still has neither a national organizational home for ecological research nor a framework for making important information widely available.[6] In the absence of government action, private groups such as the Nature Conservancy (which buys and manages land for conservation) maintain a variety of data bases consisting of biological information that is used by developers, transportation planners, and other groups. But such efforts lack the resources needed to keep information complete and current.[7]

In addition to concerns about privacy, competing interests in protecting national security and in preserving corporate secrets wall off vast sources of environmental information from public use. The huge fund of environmental information gathered by military intelligence efforts, including spy satellites with a reported resolution of a few centimeters, remains largely unavailable for civilian purposes, despite several years of work by the Central Intelligence Agency and environmental scientists to release more of it. Oil, mining, and other natural resource companies also collect and analyze a wealth of information about the earth that is protected by corporate privilege. At times companies also try to keep secret information about public risks. In 1998 the EPA's efforts to post on the Internet corporate information about possible accidents at 66,000 chemical plants around the country were countered not only by corporations but also by the Federal Bureau of Investigation and the Central Intelligence Agency, which claimed that publicizing such information might increase the chances of terrorism. An important question for the future is whether environmental information can be significantly expanded for public use while protecting legitimate competing interests.[8]

In an unfortunate coincidence, national efforts to improve environmental knowledge have been constrained just at a time of unprecedented potential for progress. Improved imaging, an ability to combine massive amounts of environmental information in layered maps using Geographic Information Systems, and rapid advances in using the Internet for instant communication have great potential to advance knowledge and inform the public more quickly than has been possible in the past.

For a country that now spends upward of $120 billion a year on pollution abatement and control, the national commitment of resources to understanding basic environmental processes, and to gathering and analyzing current data in a way that addresses crucial questions is shockingly small. Federal research and development funds deployed to advance knowledge about pollution total less than 1 percent of the annual national investment in abatement and control, and Congress shrank that amount during most of the 1980s and 1990s.[9] Environmental research and development funding as a share of federal nondefense research and development also has remained small, and declined from 6.5 percent to 6.0 percent between fiscal 1985 and fiscal 1994.[10]

The power of information, of course, is not unlimited. Information cannot make policy decisions; it can only inform them. Nor are all answerable questions worth answering. The time and money spent in gathering information are important considerations, and questions about whether additional knowledge will produce markedly better decisions are always in order. More important, policy change is inevitably slow. Even when top-quality information is available, it is often not used by policymakers, as has been noted, and findings that defy accepted wisdom may take years or decades to work their way into popular belief.

Nonetheless, no single endeavor holds as much promise for improving environmental protection as does the expansion of government and private research efforts to answer fundamental questions, to provide a basis for future planning, to provide incentives for responsible actions, and to raise the level of public debate. Such an effort has been recommended persuasively by a number of recent in-depth studies.[11] But improvements by individual agencies or organizations have limited potential since present research is widely dispersed. In the end, improving the effectiveness of information as an instrument of regulation and improving the knowledge base for policy are problems that only Congress can address.

Even with much improved information, critical questions remain to be answered. To provide an underpinning for resolving newly prominent problems, questions such as the following should be faced squarely and with optimism that competing interests can be reconciled, as they have been in the past:

—How can the now-proven national commitment to environmental protection be reconciled with the core value of protecting private property?

—Should the next policy steps rely mainly on advancing technological change, as have past approaches, or is the public prepared to support measures designed to change their daily habits?

—How can the strengths of federal and state governments be combined to produce effective policy?

116

—And how can the nation avoid the threat of hollow government, a threat characterized by the public's demand for improved environmental protection that no one is willing to pay for?

It is a mistake to concede these issues to extremists or to view them through the outdated political lens of the 1970s. The property rights issue is not a debate about whether big government is running roughshod over owners' rights. It is a debate about how and to what extent now-accepted environmental values should be integrated into land use planning. It inevitably includes the volatile question of when owners should be compensated for the loss of property value as a result of environmental regulation.

Successful accommodations between common environmental interests and private property rights are possible. The last time the issue was fully debated and successfully resolved was in the 1920s, when the Supreme Court accepted the legitimacy of zoning as a means of balancing public environmental interests and private property rights. In the 1990s pragmatic compromises have been hammered out in a piecemeal way, as noted in earlier chapters. But ad hoc deals also raise questions about fairness and equity. Only a forthright national discussion can seize this issue from extremists on both sides and rescue it from their volleys of charges and countercharges that threaten to obscure reasonable policy proposals.

Likewise, the issue of technological versus behavioral change is not about whether the federal government will keep owners from using their outdoor barbecues. It is about the degree to which voters are willing to modify their driving habits, buying choices, farming or small business practices, or homeowner decisions to improve environmental protection. Today's pragmatic politics indicates that customized requirements, financial incentives, and the new power of information may sometimes offer promising approaches. But experience also indicates that there are limits to public acceptance of habit-changing requirements. Government policies have encouraged car owners to participate in pollution inspections and to buy unleaded gas. But so far no amount of government prodding has persuaded drivers to change their commuting habits.

The issue of how to divide federal and state authority is not a question of how national policy can prevent a "race to the bottom" among states in order to minimize environmental protection. The language of the 1970s does not describe federal-state relations 30 years later and does not help policymakers approach the next generation of environmental problems. Today's debate should be about how to assign policy and administrative tasks pragmatically, in order to benefit from the strengths of each level of government. A broad consensus prevails that most air pollution and water pollution require national, and sometimes international, policies administered by state and local governments. Protecting endangered species and promoting biodiversity are inevitably national problems replete with locally unique issues. Legitimate debate can and should take place concerning federal and state responsibilities for waste disposal, safe drinking water, and other complex issues. For the most part, though, the tough questions are not whether federal or state authority should rule, but how to combine their force effectively.

The threat of hollow government has become endemic to environmental protection efforts. In a political system in which legislative promises are easily made and budget decisions are acrimonious, overcommitment has been the rule, not the exception. Passing the financial burden of federal requirements to state and local governments is a time-worn technique that does not promote environmental progress, and that may exacerbate differences between rich and poor states. Congressional attempts to break that habit—notably, restrictions on unfunded mandates—so far have not proved effective.

Adaptation is a strength of the American political system. The nation set out to build its framework of environmental policy one board at a time. Thirty years later, reformers are hammering away again. New kinds of conflicts are being resolved by tacking on new kinds of approaches. With updated understanding and better information, there is reason to be optimistic that action in the 2000s will be based not on the fears of the 1970s, but on the needs of the next century.

Notes

Introduction

1. This account of Earth Day 1970 was drawn from "Ecology: Earth Day," *Newsweek,* April 13, 1970, pp. 25–26; David Bird, "Earth Day Plans Focus on City," *New York Times,* April 20, 1970, p. 1; Gladwin Hill, "Nation Set to Observe Earth Day," *New York Times,* April 21, 1970, p. 30; Gladwin Hill, "Earth Day Goals Backed by Hickel," *New York Times,* April 22, 1970, p. 1; Gladwin Hill, "Millions Join Earth Day Observances across the Nation," and Joseph Lelyveld, "Mood Is Joyful as City Gives Its Support," *New York Times,* April 23, 1970, p. 1; Nan Robertson, "Earth's Day, like Mother's, Pulls Capital Together"; "Angry Coordinator of Earth Day: Denis Allen Hayes," *New York Times,* April 23, 1970, p. 30; "All Ages Join Pollution War in Celebration of Earth Day," *Los Angeles Times,* April 23, 1970, p. 1. The source for the statement that 20 million people participated in Earth Day events is Gaylord Nelson, "Earth Day 25 Years Later," *EPA Journal* (Winter 1995), p. 8; Jack Rosenthal, "Some Troubled by Environment Drive," *New York Times,* April 22, 1970, p. 36; David Bird, "City Announces Earth Day Plan," *New York Times,* April 17, 1970.

2. John W. Kingdon has explained that such rare events happen when there is a "coupling of problems, policy proposals, and politics," when policy windows "are opened either by the appearance of compelling problems or by happenings in the political stream." John W. Kingdon, *Agendas, Alternatives, and Public Policies,* 2d ed. (HarperCollins, 1995), pp. 19–20.

3. U.S. Environmental Protection Agency (EPA), *National Air Quality and Emissions Trends Report, 1996* (January 1998), pp. 61–65. Hazardous air pollutants are pollutants other than the six traditionally regulated by EPA. They cause especially severe health or ecological damage and include heavy metals, dioxins, and some pesticides.

4. The range of cost estimates for pollution control illustrates the difficulty of quantifying expenditures for environmental protection. The source of this estimate is Christine R. Vogan, "Pollution Abatement and Control Expenditures, 1972–94," *Survey of Current Business*, U.S. Department of Commerce, September 1996, p. 50, table 2, a data series now discontinued. For a careful analysis of various cost estimates, see J. Clarence Davies and Jan Mazurek, *Pollution Control in the United States: Evaluating the System* (Resources for the Future, 1998), pp. 103–09; and National Academy of Public Administration, *Setting Priorities, Getting Results: A New Direction for EPA* (April 1995), pp. 18–25.

5. EPA, *National Air Quality, 1996*, p. 2.

6. EPA, *National Air Quality, 1996*, pp. 3, 13; "Still Worst in U.S., California Air Is at Cleanest Level in 40 Years," *New York Times*, October 31, 1996, p. A20.

7. Davies and Mazurek, *Pollution Control in the United States*, p. 87.

8. Jesse H. Ausubel, "Industrial Ecology: A Coming of Age Story," *Resources*, Winter 1998, p. 14.

9. Detailed terms of both actions are available at the EPA's website: www.epa.gov.

10. Robert V. Percival, Alan S. Miller, Christopher H. Schroeder, and James P. Leape, *Environmental Regulation: Law, Science, and Policy* (Boston: Little, Brown, 1992), pp. 1127–31, and 1995 supplement, pp. 491–94.

11. See, for example, Michael J. Lacey, ed., *Government and Environmental Politics: Essays on Historical Developments since World War Two* (Washington, D.C.: Woodrow Wilson Center Press, 1991), an excellent group of essays on the history of environmental policy; Dyan Zaslowsky, T. H. Watkins, and the Wilderness Society, *These American Lands: Parks, Wilderness and the Public Lands* (Washington, D.C.: Island Press, 1994), a history of conservation policy concerning public lands; Robert Gottlieb, *Forcing the Spring: The Transformation of the American Environmental Movement* (Washington, D.C.: Island Press, 1993) and Philip Shabecoff, *A Fierce Green Fire: The American Environmental Movement* (New York: Hill and Wang, 1993), histories of the environmental movement; Norman J. Vig and Michael E. Kraft, eds., *Environmental Policy in the 1990s: Reform or Reaction?* 3d ed. (Congressional Quarterly Press, 1997), a group of detailed essays on specific aspects of environmental policy (see also their earlier collection, *Environmental Policy in the 1980s: Reagan's New Agenda*); Davies and Mazurek, *Pollution Control in the United States*, a detailed evaluation of federal policies; Paul R. Portney, ed., *Public Policies for Environmental Protection* (Resources for the Future, 1990), a group of essays explaining the history and provisions of federal air and water pollution, hazardous waste, and toxic pollution laws; Marc K. Landy, Marc J. Roberts, and Stephen R. Thomas, *The Environmental Protection Agency: Asking the Wrong Questions from Nixon to Clinton* (Oxford University Press, 1994), an evaluation of the performance of the Envi-

ronmental Protection Agency; DeWitt John, *Civic Environmentalism: Alternatives to Regulation in States and Communities* (Congressional Quarterly Press, 1994), a series of case studies of grass roots environmental efforts; Robert V. Percival and others, *Environmental Regulation* and supplement 1995, an excellent overview of U.S. pollution control and conservation policy.

Chapter 1

1. Sources for this account are Maryland's Blue Ribbon Citizens Pfiesteria Action Commission, *Final Report* (Annapolis, November 1997), pp. 1–17; Deborah Franklin, "The Poisoning at Pamlico Sound," *Health,* vol. 9 (September 1995), p. 108; Michael Satchell, "The Cell from Hell," *U.S. News and World Report,* July 28, 1997, p. 26; Todd Shields, "Doctors Study Effect of Pocomoke Microbe on Humans; Team Examines People Who Had Rashes, Respiratory Problems after Contact with the River," *Washington Post,* August 23, 1997, p. B8; Margaret Kriz, "Pfiesteria Hysteria," *National Journal,* September 13, 1997, p. 1783; Joby Warrick, "Microbe of Mystery; Pocomoke River Scourge Is a Versatile Killer," *Washington Post,* September 8, 1997, p. B1, and "Tiny Plants Threaten Bounty of Seas," *Washington Post,* September 23, 1997, p. A1; Paul West, "Big Pork Raises a Political Stink in Iowa; GOP Candidates Join Small Farmers against Spread of Hog Manure," *Baltimore Sun,* January 3, 1996, p. 1A; Dan Fesperman, "Bay's Economy, Future Feel Sting of Pfiesteria; Seafood Sales Are Off; Outlook Grows Dimmer with Each Incident," *Baltimore Sun,* September 21, 1997, p. 1A; John H. Cushman Jr., "Another Waterway Is Closed in Maryland; Thousands of Fish Die in Most Recent Outbreak of Microbe Problem," *New York Times,* September 15, 1997, p. A12; Richard Tapscott, "Runoff Plan in Place to Protect Bay; Farmers Would Be Paid to Erect Shoreline Buffer of Trees, Grasses in MD," *Washington Post,* October 19, 1997, p. A1; Daniel LeDuc and Peter S. Goodman, "Maryland Governor Seeks Controls to Protect Bay; State of State Speech Outlines Farming Plan," *Washington Post,* January 22, 1998, p. A1; Peter S. Goodman, "Greener Grass vs. Cleaner Water; Effort to Cure Bay Does Little about Doctoring Lawns," *Washington Post,* May 20, 1998, p. B1; Peter S. Goodman, "U.S. Unveils Plan to Help Clean Up the Bay," *Washington Post,* November 6, 1998, p. B3 (announcing federal agencies' agreement to limit runoff from federal land); Mary Hager and Larry Reibstein, "The 'Cell from Hell,'" *Newsweek,* August 25, 1997, p. 63; Peter S. Goodman, "Perdue Enters Public Debate over Pfiesteria; Poultry Industrialist Worries about Cost of Fertilizer Plan," *Washington Post,* February 11, 1998, p. B4; Todd Shields, "For Watermen, It Was Never a Fish Story; Pocomoke's Small Group Overcame Large Hurdle: Proving They Were on Deadly Ground," *Washington Post,* September 15, 1997, p. C1; Associated Press, "A Chesapeake Parasite Is Killing Fish and Making People Ill," *New York Times,* September 7, 1997, p. 41; Dail Willis and Marcia Myers, "Md. to Reopen Pocomoke Tomorrow; Toxins That Killed Fish Are Gone, Officials Believe; Federal Aid Allocated; Senators Tour Area; Residents Offer Ideas on Dealing with Issue," *Baltimore Sun,* August 12, 1997, p. 1B; Dan Fesperman and Marcia Myers, "New Urgency in Hunt for Cause of

Fish Ills; Kill Area Continues to Spread North along Pocomoke; Anglers Seek Evaluation of Threat to Human Health," *Baltimore Sun,* August 29, 1997, p. 1A; Rob Hiaasen, "Trouble in the Water," *Baltimore Sun,* August 30, 1997, p. 1D; Margaret Kriz, "Fish and Foul," *National Journal,* February 28, 1998, p. 450.

2. Executive Office of the President, Council on Environmental Quality, *Environmental Quality: The Twenty-fifth Anniversary Report of the Council on Environmental Quality* (Government Printing Office, 1997), p. 242.

3. U.S. Environmental Protection Agency (EPA), *Clean Water Action Plan: Restoring and Protecting America's Waters,* Overview, February 14, 1998, p. i (revised August 10, 1998).

4. EPA, *Clean Water Action Plan,* p. 10.

5. Council on Environmental Quality, *Environmental Quality,* pp. 249–50.

6. EPA, *Report on 1994 National Water Quality Inventory,* p. 12; Council on Environmental Quality, *Environmental Quality,* pp. 12–14.

7. The cause of the Milwaukee outbreak was variously reported as runoff from dairy farms and sewage from a nearby treatment plant or leaking septic tanks. Robert V. Percival, Alan S. Miller, Christopher H. Schroeder, and James P. Leape, *Environmental Regulation: Law, Science, and Policy* (Boston: Little, Brown, 1995 supplement), p. 348; Sara Terry, "Drinking Water Comes to a Boil," *New York Times,* September 26, 1993, sec. 6, p. 42. The number of people affected is from *Budget of the United States Government, Fiscal Year 2000* (GPO 1999), p. 96.

8. Josh White and Maria Glod, "Cost of Replacing Underground Tanks Sinks Some Gas Stations," *Washington Post,* January 4, 1999, p. B1.

9. Council on Environmental Quality, *Environmental Quality,* p. 9.

10. U.S. Geological Survey, "Estimated Use of Water in the United States in 1995," USGS circular 1200 (October 1998).

11. Toxic releases for 1995 are reported in EPA, *1995 Toxics Release Inventory: Public Data Release* (1997), as summarized in J. Clarence Davies and Jan Mazurek, *Pollution Control in the United States: Evaluating the System* (Washington, D.C.: Resources for the Future, 1998), pp. 86–88; Council on Environmental Quality, *Environmental Quality,* pp. 231–32.

12. Council on Environmental Quality, *Environmental Quality,* pp. 93–94, 183.

13. EPA, Regional Human Health Risk Rankings, summary sheet, 1993 (obtained from EPA Regional and State Planning Division); Science Advisory Board, *Reducing Risk: Setting Priorities and Strategies for Environmental Protection* (EPA, September 1990), p. 14.

14. Council on Environmental Quality, *Environmental Quality,* pp. 93–94.

15. See Wayne R. Ott and John W. Roberts, "Everyday Exposure to Toxic Pollutants," *Scientific American,* February 1998, pp. 86–91.

16. Ott and Roberts, "Everyday Exposure to Toxic Pollutants"; Council on Environmental Quality, *Environmental Quality,* pp. 92–94 (on secondhand smoke); Denise Scheberle, *Federalism and Environmental Policy: Trust and the Politics of Implementation* (Georgetown University Press, 1997), pp. 102–21; Warren E. Leary, "Radon More Dangerous in Air than in Water," *New York Times,* September 16, 1998, p. A13.

17. EPA, *National Air Quality and Emissions Trend Report, 1995* (October 1996), pp. 1–4; EPA, Office of Air Quality Planning and Standards, *National Air Quality and Emissions Trend Report, 1996* (January 1998), pp. 2, 3, 21, 30, 34.

18. EPA, *National Air Quality, 1995*, pp. 21–25; EPA, *National Air Quality, 1996*, pp. 21–25, table A-4; Council on Environmental Quality, *Environmental Quality*, pp. 179–86.

19. EPA, *National Air Quality, 1996*, pp. 31, 34, and table A-6.

20. EPA, *National Air Quality, 1995*, pp. 49–53; Council on Environmental Quality, *Environmental Quality*, pp. 188–89.

21. John H. Cushman Jr., "E.P.A. to Hunt Dangers in Everyday Products," *New York Times*, August 31, 1998, p. A1.

22. Remarks by Arthur R. Sigel, Chemical Manufacturers Association, National Press Club, Washington, D.C., January 27, 1999. Companies promised to post all data on the Chemical Manufacturers Association website (www.cmahq.org).

23. Council on Environmental Quality, *Environmental Quality*, pp. 11–12, 179-86; EPA, *National Air Quality, 1995*, pp. 18–25; EPA, *National Air Quality, 1996*, pp. 9–10, 63; National Research Council, *Rethinking the Ozone Problem in Urban and Regional Air Pollution* (National Academy Press, 1991), pp. 283–84, 301. The National Research Council report also points out that volatile organic compounds produced by vegetation may contribute to the atmosphere as much as half of the compounds that are elements of ozone pollution (pp. 8, 302); John H. Cushman Jr., "Record Penalty Likely against Diesel Makers: $1 Billion Deal Cites Clean-Air Violations," *New York Times*, October 22, 1998, p. A1.

24. EPA, *National Air Quality, 1996*, p. 2, tables A-2 through A-8; Davies and Mazurek, *Pollution Control in the United States*, pp. 58, 81, 87.

25. The President's Council on Environmental Quality reports that more than two-thirds of the value of the country's $40 billion fishing and fish-processing industry depends on coastal wetlands, for example. Council on Environmental Quality, *Environmental Quality*, pp. 270–71, 304–05 (forests reduce carbon dioxide); L. E. Drinkwater and others, "Legume-Based Cropping Systems Have Reduced Carbon and Nitrogen Losses," *Nature*, November 19, 1998, p. 262 (full text available at www.nature.com).

26. Council on Environmental Quality, *Environmental Quality*, pp. 129–47. Information about the Department of the Interior's effort to promote habitat conservation plans is located on the department's website, www.doi.gov. Federal officials and private groups are also working together to create a report card on the health of the nation's ecosystems, an effort coordinated by the Heinz Center for Science, Economy, and the Environment, www.heinzctr.org. The World Wildlife Fund is attempting to rank eco-regions in the United States according to the severity of threats to them, www.wwf.org.

27. Council on Environmental Quality, *Environmental Quality*, p. 140 (on the California debate); Louis Jacobson, "A Green City Gets a Comeuppance," *National Journal*, September 12, 1998, p. 2119 (on the Portland debate).

28. Because the federal government acquired vast public lands as a result of the Louisiana Purchase and agreements by eastern states to give up claims to western lands, most public lands are located in the West. Eighty-three percent of lands in Nevada and 28 percent of lands in Montana, for example, are owned by the federal government. Council on Environmental Quality, *Environmental Quality*, pp. 171–72.

29. Council on Environmental Quality, *Environmental Quality*, pp. x, 274.

30. Science Advisory Board, *Reducing Risk*, p. 13.

31. Council on Environmental Quality, *Environmental Quality*, pp. 71, 171, 281, 466 (table 50); EPA, *National Air Quality, 1996*, p. 2.

32. Council on Environmental Quality, *Environmental Quality*, pp. x, 73, 169–72 (on land ownership), 245–67 (on coastal issues); Department of Commerce, *Statistical Abstract of the United States, 1997*, table 370, p. 229 (on developed land).

33. Richard Hofstadter, *The Age of Reform: From Bryan to F.D.R.* (New York: Knopf, 1955), p. 116.

34. U.S. Department of Commerce, Bureau of the Census, "Large Farms Are Thriving in the United States," Agricultural Brief AB/96-1 (July 1996); David E. Ervin and others, "Agriculture and Environment: A New Strategic Vision," *Environment*, July 17, 1998, p. 8.

35. Marian R. Chertow and Daniel C. Esty, eds., *Thinking Ecologically: The Next Generation of Environmental Policy* (Yale University Press, 1997), p. 201 (most farms not subject to Clean Water Act); Environmental Law Institute, *Enforceable State Mechanisms for the Control of Nonpoint Source Water Pollution* (October 1997), p. iii (on state agricultural exemptions).

36. Several times, Congress urged states to take action. In 1987 Congress added a provision to the Clean Water Act to encourage states to control runoff that flowed through storm sewers. In 1990 Congress added a provision to the Coastal Zone Management Act to require states with coastal plans to include programs to control runoff. Percival and others, *Environmental Regulation*, pp. 944–49. A 1997 study found scattered state provisions in forestry, fish and game, nuisance, water pollution, and land use laws. Environmental Law Institute, *Enforceable State Mechanisms*, p. i.

37. William H. Rodgers Jr., *Environmental Law* (St. Paul, Minn.: West, 1977), pp. 301–03.

38. Davies and Mazurek, *Pollution Control*, p. 109.

39. For a discussion of national and state interests in these issues, see chapter 4.

40. The shortcomings of such categorization are now widely appreciated, and federal and state governments have taken steps to build bridges among categories.

41. A. Myrick Freeman III, "Water Pollution Policy," in Paul R. Portney, ed., *Public Policies for Environmental Protection* (Resources for the Future, 1990), p. 143.

42. James T. Patterson, *Grand Expectations: The United States, 1945–1974* (Oxford University Press, 1996), pp. 70–72.

Chapter 2

1. Warren Weaver, Jr., "Hickel Reverses Coast Oil Stand," *New York Times,* February 8, 1969, p. A1; Gladwin Hill, "Santa Barbara Harbor Closed; Oil Fouls Beaches, Fire Hazard Feared," *New York Times,* February 6, 1969, p. A1; Robert B. Semple, Jr., "Nixon Gives Truman That Old Piano of His, and All Is Harmony," *New York Times,* March 22, 1969, p. A1 (includes a description of the president's aerial tour of the blowout); "Head of Union Oil Denies Remarks on Loss of Birds in Slick," *New York Times,* February 19, 1969, p. 32 (Hartley quotation); Warren Weaver, Jr., "A Request by Hickel Stops Drilling off Santa Barbara," *New York Times,* February 4, 1969, p. A1; Warren Weaver, Jr., "President May Call G.I.'s to Combat Coast Oil Slick," *New York Times,* February 7, 1969, p. A1; *San Francisco Examiner,* January 30, 1969, p. 1 ("the accident that experts said wouldn't happen"), and February 12, 1969; Executive Office of the President's Council on Environmental Quality, *Environmental Quality: the Twenty-fifth Anniversary Report of the Council on Environmental Quality* (GPO, 1997), p. 7 (which points out that the federal government failed to protect ecological values); Gladwin Hill, "Santa Barbarans Wait for Oil 'Powder Keg' to Blow Up," *New York Times,* May 30, 1969, p. A10.

2. "Oil in L.I. Sound Imperiling Ducks; Inquiry Under Way," *New York Times,* January 21, 1969, p. 49.

3. "Oil Slicks off Connecticut Threaten Resort Areas," *New York Times,* May 8, 1969, p. 49.

4. "Leaking Barge Spreads Oil for 15 Miles on Mississippi," *New York Times,* June 19, 1969, p. 90.

5. "Oil Barge Aground; Bay State Periled," *New York Times,* September 17, 1969, p. 93.

6. "Oil Slick Fire Damages 2 River Spans," *Cleveland Plain Dealer,* June 23, 1969, p. 11C; "Cleveland: Where the River Burns," editorial, *Cleveland Plain Dealer,* June 24, 1969, p. 10-A; Jack A. Seamonds, "In Cleveland, Clean Waters Give New Breath of Life," *U.S. News & World Report,* June 18, 1984, p. 68; Council on Environmental Quality, *Environmental Quality,* p. 7.

7. U.S. Senate Committee on Public Works, "National Air Quality Standards Act of 1970," S. Rept. 91-1196 (GPO, September 17, 1970), p. 2.

8. S. Rept. 91-1196, p. 37.

9. Theodore H. White, *The Making of the President 1972* (New York: Atheneum, 1973), pp. 75–82; David Nevin, *Muskie of Maine* (Random House, 1972); p. 117; Marc K. Landy, Marc J. Roberts, and Stephen R. Thomas, *The Environmental Protection Agency: Asking the Wrong Questions from Nixon to Clinton,* expanded ed. (Oxford University Press, 1994), pp. 26–33; J. Clarence Davies III and Barbara S. Davies, *The Politics of Pollution,* 2d ed. (Indianapolis: Pegasus, 1975), pp. 49–56; Alfred A. Marcus, *Promise and Performance: Choosing and Implementing an Environmental Policy* (Westport, Conn.: Greenwood Press, 1980), p. 60.

10. White, *The Making of the President 1972,* p. 76; Nevin, *Muskie of Maine,* pp. 13, 18, 19.

11. *Vanishing Air,* quoted in Charles McCarry, *Citizen Nader* (New York: New American Library, 1973), pp. 170–73.

12. Marcus, *Promise and Performance,* pp. 53–62.

13. Stephen E. Ambrose, *Nixon: The Triumph of a Politician, 1962–1972,* vol. 2 (Simon and Schuster, 1989), pp. 395–96; White, *The Making of the President 1972,* pp. 52–53.

14. Ray Price quoted in Ambrose, *Nixon: The Triumph,* p. 397.

15. White, *The Making of the President 1972,* p. 55; Robert S. Diamond, "What Business Thinks," *Fortune,* February 1970, p. 118.

16. Ambrose, *Nixon: The Triumph,* p. 413 (on the Harris poll); Nevin, *Muskie of Maine,* p. 19 (on the poll results); Marcus, *Promise and Performance,* pp. 31–49.

17. Michael J. Lacey, ed., *Government and Environmental Politics: Essays on Historical Developments since World War Two* (Washington, D.C.: Woodrow Wilson Press, 1991), pp. 82–99.

18. Paul Starr, *The Social Transformation of American Medicine* (Basic Books, 1982), pp. 180–85, 342.

19. James T. Patterson, *Grand Expectations: The United States, 1945–1974* (Oxford University Press, 1996), pp. 712–13; John Brooks, *The Great Leap: The Past Twenty-Five Years in America* (Harper and Row, 1966), pp. 260–61.

20. Donald L. Miller, *City of the Century: The Epic of Chicago and the Making of America* (Simon and Schuster, 1996), p. 424.

21. Charles O. Jones, *Clean Air: The Policies and Politics of Pollution Control* (University of Pittsburgh Press, 1975), pp. 20–41; Davies and Davies, *The Politics of Pollution,* pp. 19–20, 158–60; Matthew A. Crenson, *The Un-Politics of Air Pollution: A Study of Non-Decisionmaking in the Cities* (Johns Hopkins University Press, 1971), p. 11.

22. Davies and Davies, *The Politics of Pollution,* pp. 47–50.

23. Robert V. Percival, Alan S. Miller, Christopher H. Schroeder, and James P. Leape, *Environmental Regulation: Law, Science, and Policy* (Boston: Little, Brown), p. 104.

24. Percival and others, *Environmental Regulation,* pp. 873–75; Daniel M. Rohrer, David C. Montgomery, Mary E. Montgomery, David J. Eaton, and Mark G. Arnold, *The Environment Crisis: A Basic Overview of the Problem of Pollution* (Skokie, Ill.: National Textbook, 1970), p. 221.

25. Davies and Davies, *The Politics of Pollution,* pp. 29–36; Norman J. Vig and Michael E. Kraft, eds., *Environmental Policy in the 1990s: Toward a New Agenda* (Congressional Quarterly Press, 1990), pp. 9–11; Percival and others, *Environmental Regulation,* pp. 103–06.

26. Davies and Davies, *The Politics of Pollution,* pp. 32–35.

27. Lacey, *Government and Environmental Politics,* pp. 23, 82–84, 117–18, 148, 152. An excellent history of the development of the early conservation movement is Dyan Zaslowsky, T. H. Watkins, and the Wilderness Society, *These American Lands: Parks, Wilderness and the Public Lands* (Washington, D.C.: Island Press), 1994.

28. Percival, *Environmental Regulation,* p. 104.

29. Lacey, *Government and Environmental Politics*, pp. 36–37, 117–28, 152; Patterson, *Grand Expectations*, pp. 725–26; Percival and others, *Environmental Regulation*, pp. 104–07.

30. Paul R. Portney, "Air Pollution Policy," in Paul R. Portney, ed., *Public Policies for Environmental Protection* (Resources for the Future, 1990), p. 48, table 3-4, and personal communication with the author.

31. Marcus, *Promise and Performance*, p. 19.

32. Davies and Davies, *The Politics of Pollution*, p. 90; Patterson, *Grand Expectations*, p. 725 (on the membership of environmental groups).

33. Lacey, *Government and Environmental Politics*, pp. 95–99.

34. Patterson, *Grand Expectations*, p. 727 (on Earth Day); Robert Cameron Mitchell, "Public Opinion and Environmental Politics in the 1970s and 1980s," in Norman J. Vig and Michael E. Kraft, eds., *Environmental Policy in the 1980s: Reagan's New Agenda* (Congressional Quarterly Press, 1984), pp. 51–52 (a majority supported national action).

35. White, *The Making of the President 1972*, p. 45.

36. Patterson, *Grand Expectations*, pp. 725–26; Edward C. Banfield and James Q. Wilson, *City Politics* (Harvard University Press, 1963), pp. 78–79, 330–46; Crenson, *The Un-Politics of Air Pollution*, p. 17.

37. Patterson, *Grand Expectations*, p. 565.

38. Alfred Marcus, "The Environmental Protection Agency," in James Q. Wilson, ed., *The Politics of Regulation* (Basic Books, 1980), p. 273; Percival and others, *Environmental Regulation*, pp. 1–6 (Reagan quoted on p. 5).

39. Council on Environmental Quality, *Environmental Quality*, pp. 6–7.

40. Percival and others, *Environmental Regulation*, pp. 1–6. The seemingly sudden increase in scientific findings about environmental problems was no coincidence. Growing concern about the effects of industrial chemicals and auto pollution led Congress to fund considerable research, including a five-year study of air pollution effects in 1955 and an inquiry into the health effects of auto pollution in 1960 (pp. 104–06).

41. Patterson, *Grand Expectations*, pp. 726–27.

42. Davies and Davies, *The Politics of Pollution*, pp. 19–20; Lacey, *Government and Environmental Politics*, pp. 36–42.

43. Percival and others, *Environmental Regulation*, pp. 725–27; Lacey, *Government and Environmental Politics*, p. 100–02.

44. Lacey, *Government and Environmental Politics*, pp. 92–95, 102–06.

45. Lacey, *Government and Environmental Politics*, p. 89; Percival and others, *Environmental Regulation*, pp. 1–6; Patterson, *Grand Expectations*, p. 565.

46. Patterson, *Grand Expectations*, pp. 447, 562–92. This optimistic time also produced a handful of environmental initiatives, some of which encouraged state action on air pollution and set national motor vehicle standards, created national parks and wilderness areas, and endeavored to stop the proliferation of billboard advertising. But these measures were overshadowed by more momentous national initiatives (p. 726).

47. Patterson, *Grand Expectations*, p. 595, 637–52, 722. For example, the use of defoliants such as Agent Orange destroyed half the forests in South Vietnam.

48. Patterson, *Grand Expectations*, pp. 637–90, 708; White, *The Making of the President 1972*, p. 39.

49. Theodore J. Lowi, *The End of Liberalism: Ideology, Policy and the Crisis of Public Authority* (W. W. Norton, 1969); Kenneth Culp Davis, *Administrative Law Text*, 3d ed. (St. Paul, Minn.: West, 1972), p. 88 (quoted passage is on p. 92).

50. Michael Barone, *Our Country: The Shaping of America from Roosevelt to Reagan* (New York: The Free Press, 1990), pp. 273, 354–55.

51. Testimony before the House Committee on Science and Astronautics, Subcommittee on Science, Research, and Development, quoted in Crenson, *The Un-Politics of Air Pollution*, p. 5.

52. Crenson, *The Un-Politics of Air Pollution*, pp. 5 ("municipal inaction . . ."), 87–91, 159–65, 184.

53. Davies and Davies, *The Politics of Pollution*, pp. 164–66.

54. Elizabeth H. Haskell and Victoria S. Price, *State Environmental Management: Case Studies of Nine States* (Praeger, 1973), pp. 244–45.

55. Haskell and Price, *State Environmental Management*, pp. 244–45.

56. Lacey, *Government and Environmental Politics*, p. 47.

57. Rohrer and others, *The Environment Crisis*, pp. 224–25.

58. President Richard M. Nixon, "Environmental Quality," *Weekly Compilation of Presidential Documents*, vol. 6, no. 7, February 16, 1970, p. 166.

59. House Report 91-1146, in *Legislative History of the Clean Air Act*, 1970, p. 3. In the 1990s, the idea of "a race to the bottom" is too simplistic to describe state interests in environmental protection (see chapter 3 of this volume), but the issue is still debated. Richard L. Revesz, professor of law at New York University Law School, has made a cogent theoretical argument against such an economic imperative. See his "Rehabilitating Interstate Competition: Rethinking the 'Race-to-the-Bottom' Rationale for Federal Environmental Regulation," *New York University Law Review*, vol. 67 (December 1992).

60. Daphne A. Kenyon and John Kincaid, eds., *Competition among State and Local Governments* (Washington, D.C.: Urban Institute Press, 1991), p. 223; Peter K. Eisinger, *The Rise of the Entrepreneurial State: State and Local Economic Development Policy in the United States* (University of Wisconsin Press, 1988), pp. 16, 153.

61. Patterson, *Grand Expectations*, pp. 565–68 (on redistricting effects on states); Haskell and Price, *State Environmental Management*, pp. 243–44 (on public attitudes in states); Rohrer and others, *The Environment Crisis*, pp. 145–50 (on changes in state air pollution control programs); Marcus, *Promise and Performance*, p. 67 (on California auto pollution regulation); Robert S. Diamond, "What Business Thinks," *Fortune*, February, 1970, p. 118.

62. Robert J. Samuelson, *The Good Life and Its Discontents: The American Dream in the Age of Entitlement 1945–1995* (Random House, 1995), pp. 7–8, 34–41. Patterson, *Grand Expectations*, pp. 313–26.

63. Alfred D. Chandler Jr., *The Visible Hand: The Managerial Revolution in American Business*, pp. 480–83.

64. Wilson, *The Politics of Regulation*, pp. 365–66 ("the perceived legitimacy of business enterprise has declined"); Chandler, *The Visible Hand*, pp. 480–82 (conglomerates). With regard to business resistance to government requests for information, see, for example, Alfred Marcus, "Environmental Protection Agency," in Wilson, *The Politics of Regulation*, pp. 282–83.

65. McCarry, *Citizen Nader*, pp. 13–34 (on General Motors investigating Nader); William H. Rodgers Jr., *Environmental Law* (West, 1977), p. 288 (on federal suit against auto manufacturers); Council on Environmental Quality, *Environmental Quality*, p. 7 (on the chemical companies that attacked *Silent Spring*).

66. Chandler, *The Visible Hand*, pp. 10–11, 480–83; Brooks, *The Great Leap*, pp. 38–55.

67. Clean Air Act Amendments of 1970, P.L. 91-604; Rodgers, *Environmental Law*, pp. 217–67.

68. Rodgers, *Environmental Law*, pp. 359–96 (on the Water Pollution Control Act); Marcus, "Environmental Protection Agency," pp. 274–76.

69. Endangered Species Act of 1973, P.L. 93-205.

70. Federal Environmental Pesticide Control Act of 1972, P.L. 92-516.

71. Marine Protection, Research, and Sanctuaries Act of 1972, P.L. 92-532.

72. Lacey, *Government and Environmental Politics*, pp. 195–99.

73. Construction grants to improve solid waste disposal were included in the Resource Recovery Act of 1970, P.L. 91-512.

Chapter 3

1. The phrase "hollow government" was first used as the title of an article in *Government Executive* (October 1989), by Mark L. Goodstein, referred to by David S. Broder in his column "Fumble, Punt, Fumble," *Washington Post*, November 1, 1989, p. A25.

2. James T. Patterson, *Grand Expectations: The United States, 1945–1974* (Oxford University Press, 1996), pp. 740–41; Alfred A. Marcus, *Promise and Performance: Choosing and Implementing an Environmental Policy* (Westport, Conn.: Greenwood Press, 1980), pp. 124–31; Stephen E. Ambrose, *Nixon: The Triumph of a Politician, 1962–1972* (Simon and Schuster, 1989), p. 292, quotation on p. 460.

3. Patterson, *Grand Expectations*, pp. 759–65; Theodore H. White, *The Making of the President 1972* (New York: Atheneum, 1973), pp. 80–82.

4. Daniel Yergin, *The Prize: The Epic Quest for Oil, Money, and Power* (Simon and Schuster, 1991), pp. 604–09; Patterson, *Grand Expectations*, pp. 783–86.

5. Marcus, *Promise and Performance*, pp. 129–30.

6. Safe Drinking Water Act, P.L. 93-583.

7. Examples are the Federal Land Policy and Management Act of 1976, P.L. 94-579, and the National Forest Management Act of 1976, P.L. 94-588.

8. Robert V. Percival, Alan S. Miller, Christopher H. Schroeder, and James P. Leape, *Environmental Regulation: Law, Science, and Policy* (Little, Brown, 1992 and 1995 supplements), p. 288.

9. Alice M. Rivlin, *Reviving the American Dream: The Economy, the States, and the Federal Government* (Brookings, 1992), pp. 57–58.

10. Everett Carll Ladd and Karlyn H. Bowman, *Attitudes toward the Environment* (Washington, D.C.: American Enterprise Institute for Public Policy Research, 1995), pp. 1–25.

11. Ladd and Bowman, *Attitudes toward the Environment*, pp. 10, 33.

12. Christine R. Vogan, "Pollution Abatement and Control Expenditures, 1972–94," *Survey of Current Business*, U.S. Department of Commerce, September 1996, pp. 48–49.

13. Clean Air Act Amendments of 1990, P.L. 101-549; Percival, *Environmental Regulation*, p. 766.

14. Ladd and Bowman, *Attitudes toward the Environment*, pp. 2–47. The President's Council on Environmental Quality review of polling data also found strong public support for further government action. In 1994, 81 percent of those polled thought environmental regulations were about right or did not go far enough, up from two-thirds in 1973; Executive Office of the President's Council on Environmental Quality, *Environmental Quality: The Twenty-fifth Anniversary Report of the Council on Environmental Quality* (GPO, 1997), p. 25; Alan Riding, "New Catechism for Catholics Defines Sins of Modern World," *New York Times*, November 17, 1992.

15. Joby Warrick, "Opponents Await Proposal to Limit Air Particulates," *Washington Post*, November 27, 1996, p. A1.

16. Samuel H. Beer, *To Make a Nation* (Harvard University Press, 1993), p. 2.

17. John McCain, "Nature Is Not a Liberal Plot," *New York Times*, November 22, 1996, p. A31.

18. Robert Cameron Mitchell, "Public Opinion and Environmental Politics in the 1970s and 1980s," in Norman J. Vig and Michael E. Kraft, eds., *Environmental Policy in the 1980s* (Congressional Quarterly Press, 1984), pp. 51–71 (on the Reagan administration's environmental policy); Michael E. Kraft, "Environmental Policy in Congress: Revolution, Reform, or Gridlock?" in Vig and Kraft, eds., *Environmental Policy in the 1990s*, pp. 128–38 (on actions of the 104th Congress).

19. Executive Order 12898, February 11, 1994. For one call for policy reform, see Robert D. Bullard, *Dumping in Dixie: Race, Class, and Environmental Quality* (Boulder, Colo.: Westview Press, 1990). For suggested redirection of environmental justice efforts, see Christopher H. Foreman Jr., *The Promise and Peril of Environmental Justice* (Brookings, 1998).

20. David M. Halbfinger, "6-Year Fight against Hospital Incinerator Pays Off," *New York Times*, June 30, 1997, p. B3; Margaret Kriz, "The Color of Poison," *National Journal*, July 11, 1998, p. 1608.

21. *Pennsylvania Coal Co.* v. *Mahon*, 260 U.S. 393 (1922).

22. *Euclid* v. *Ambler Realty Co.*, 272 U.S. 365 (1926).

23. *Penn Central Transportation Co.* v. *New York City*, 438 U.S. 104 (1978).

24. *Lucas* v. *South Carolina Coastal Council,* 505 U.S. 1003 (1992).

25. *Nollan* v. *California Coastal Commission,* 483 U.S. 825 (1987); and *Dolan* v. *City of Tigard,* 512 U.S. 374 (1994).

26. Michael E. Kraft, "Environmental Policy in Congress: Revolution, Reform, or Gridlock?" in Vig and Kraft, eds., *Environmental Policy in the 1990s,* p. 130. Percival and others, *Environmental Regulation,* 1995 supplement, p. 435.

27. Lettie McSpadden, "Environmental Policy in the Courts," in Vig and Kraft, *Environmental Policy in the 1990s,* p. 174; Margaret Kriz, "Taking Issue," *National Journal,* June 1, 1996, p. 1200. Kriz notes that Arizona and Washington state voters overturned state property-rights laws on p. 1201.

28. *Lucas* v. *South Carolina Coastal Council,* 505 U.S. 1003 (1992).

29. Jeff McLaughlin, "In Nature's Defense: Halting River Pollution Conflicts with Property Rights," *Boston Globe,* June 1, 1997, Northwest Weekly, p. 1.

30. Arnold M. Howitt and Alan Altshuler, "The Politics of Controlling Auto Air Pollution," in José A. Gómez-Ibáñez, William B. Tye, and Clifford Winston, eds., *Essays in Transportation Economics and Policy: A Handbook in Honor of John R. Meyer* (Brookings 1999), pp. 237–46; Arnold M. Howitt, *Managing Federalism: Studies in Intergovernmental Relations* (Congressional Quarterly Press, 1984), pp. 122–55. See also U.S. Congressional Budget Office, *Federalism and Environmental Protection: Case Studies for Drinking Water and Ground-Level Ozone,* November 1997, pp. 50–51.

31. Ladd and Bowman, *Attitudes toward the Environment,* pp. 32–34.

32. Ladd and Bowman, *Attitudes toward the Environment,* pp. 28–32; the quotation appears on p. 32.

33. U.S. Department of Commerce, "Pollution Abatement," pp. 48–52.

34. Robert Gottlieb, *Forcing the Spring: The Transformation of the American Environmental Movement* (Washington, D.C.: Island Press, 1993), pp. 136–40.

35. Personal communication, Douglas Hall, vice-president, Nature Conservancy.

36. Information about the Alliance for Environmental Innovation is located at http://www.edf.org/alliance. Information about efforts by the World Wildlife Fund is located at http://www.wwf.org.

37. Tom Kenworthy, "Conservationists Challenge Ranchers' Hold on State Lands," *Washington Post,* September 9, 1997, p. A1.

38. Jeffrey Krasner, "State to Get Car Insurer with a Cause," *Wall Street Journal,* January 28, 1998, p. NE1.

39. Dick Kirschten, "Green Grows the Immigration Debate," *National Journal,* March 7, 1998, p. 532.

40. David Vogel, "International Trade and Environmental Regulation," in Vig and Kraft, *Environmental Policy in the 1990s,* p. 356.

41. J. Clarence Davies and Jan Mazurek, *Pollution Control in the United States* (Resources for the Future, 1998), pp. 12–13; Percival and others, *Environmental Regulation,* pp. 766–68.

42. U.S. Environmental Protection Agency, *National Air Quality and Emissions Trends Report, 1995* (GPO, 1995), pp. 51–53.

43. Council on Environmental Quality, *Environmental Quality,* p. 91.

44. EPA, "Safe Drinking Water Act Amendments of 1996," General Guide to Provisions, 1996; Council on Environmental Quality, *Environmental Quality*, pp. 186–87, 240.

45. Davies and Mazurek, *Pollution Control*, p. 14.

46. The *Registry of International Treaties and Other Agreements* (United Nations Environment Program, 1996) listed 208 mulilateral agreements.

47. Montreal Protocol on Substances that Deplete the Ozone Layer (1987); London Amendments to Montreal Protocol (1990).

48. Report of the Appellate Body, United States—Import Prohibition of Certain Shrimp and Shrimp Products, World Trade Organization, October 12, 1998; Report of the Panel, United States—Restrictions on Imports of Tuna, GATT Doc. DS21/R, August 16, 1991.

49. Executive Office of the President, *Budget of the United States Government, Fiscal Year 1999*, p. 80; "The Sting in Trade's Tail," *Economist*, April 18, 1998, p. 70; Abram Chayes and Antonia Handler Chayes, *The New Sovereignty: Compliance with International Regulatory Agreements* (Harvard University Press, 1995), pp. 14–16, 184–90.

50. Vig and Kraft, *Environmental Policy in the 1990s*, p. 18.

51. Council on Environmental Quality, *Environmental Quality*, pp. 9, 236.

52. Karen Marshall, "State Environmental and Natural Resources Expenditures," in *Resource Guide to State Environmental Management*, 4th ed. (Council of State Governments, 1996), p. 127.

53. An analysis of the law's first year of operation by the Congressional Budget Office concluded that cost estimates had improved under its measures, but the narrow definition of mandates and the ease with which the requirement could be avoided placed its long term impact in doubt. Executive Order 12875, October 1993; Unfunded Mandates Reform Act of 1995, P.L.104-4; Theresa A. Gullo and Janet M. Kelly, "Federal Unfunded Mandate Reform: A First Year Retrospective," *Public Administration Review*, September/October 1998, p. 379.

54. *New York v. United States*, 505 U.S. 144 (1992).

55. *United States v. Lopez*, 514 U.S. 549 (1995).

56. *City of Boerne v. Flores*, 521 U.S. 507 (1997).

57. Information about the Natural Heritage programs can be found at the Nature Conservancy's website, http://www.tnc.org.

58. Percival and others, *Environmental Regulation*, p. 821.

59. Percival and others, *Environmental Regulation*, pp. 818–21; Richard L. Revesz, "Federalism and Interstate Environmental Externalities," *University of Pennsylvania Law Review*, vol. 144 (June 1996), pp. 2349–61; Joby Warrick, "EPA Orders Emission Reductions," *Washington Post*, September 25, 1998, p. A4.

60. Mike Allen, "Connecticut Joins Lawsuit over Pollution in Sound," *New York Times*, March 24, 1998, p. A24.

61. Victor S. Rezendes, "Environmental Protection: Issues Facing the Energy and Defense Environmental Management Programs," GAO testimony before

House Subcommittees on Military Procurement and Military Readiness, March 21, 1996; *Budget of the United States Government, Fiscal Year 2000* (1999), p. 84.

62. Percival and others, *Environmental Regulation*, p. 766. Paul R. Portney, ed., *Public Policies for Environmental Protection* (Resources for the Future, 1990), pp. 7–25, includes a lucid summary of basic principles and tensions in U.S. policy.

63. Davies and Mazurek, *Pollution Control*, pp. 16–19; Vig and Kraft, *Environmental Policy in the 1990s*, pp. 4–7.

64. William D. Ruckelshaus, "Stepping Stones," *Environmental Forum*, March/April 1998, p. 36.

65. "Oklahoma Governor Adopts Animal Waste Task Force Advice," *State Environmental Monitor*, January 12, 1998, pp. 11–12.

66. Daniel Roth, "The Ray Kroc of Pigsties," *Forbes 400*, October 13, 1997, p. 120.

67. In this context, prosperity means more than simply economic growth. It refers to the building and maintenance of physical and social infrastructure that allows the state and its residents to flourish.

68. Adam B. Jaffe, Steven R. Peterson, Paul R. Portney, and Robert N. Stavins, "Environmental Regulation and the Competitiveness of U.S. Manufacturing: What Does the Evidence Tell Us?" *Journal of Economic Literature*, March 1995, p. 141; Alan Carlin, *Environmental Investments: The Cost of a Clean Environment, A Summary* (U.S. Environmental Protection Agency, 1990), pp. 2–6. For corporate surveys, see, for example, "Where Will Facilities Go," *Area Development* annual survey, December, 1996, a survey primarily of manufacturing companies. Environmental regulations ranked seventh, after labor costs, highway accessibility, construction costs, energy availability and costs, skilled labor, and state and local incentives. Personal communication, James A. Schriner, vice president, Fantus Consulting (a firm that advises businesses on relocation decisions), 1996.

69. Keon S. Chi and Drew Leatherby, *State Business Incentives: Trends and Options for the Future* (Council of State Governments, 1997), p. 8.

70. Arik Levinson, "Environmental Regulations and Manufacturers' Location Choices: Evidence from the Census of Manufactures, *Journal of Public Economics*, vol. 62 (October 1996), pp. 6, 5–29. Levinson, a professor of economics at the University of Wisconsin, analyzed Census of Manufactures data and a wide variety of state environmental indicators and found that industries that spend more on pollution abatement do not seem to avoid states with strict environmental rules. On the other hand, Levinson and other economists have found some signs that strict environmental rules may exert at least a limited influence on business location in particular situations, including small business start-ups and branch plant placement; Jaffe and others, "Environmental Regulation," p. 149.

71. Organization for Economic Cooperation and Development (OECD), *Integrating Environment and Economy: Progress in the 1990s* (1996), pp. 9, 41. "Empirical research on these questions suggests that businesses do not generally

relocate on the basis of any additional economic burdens that may be imposed by environmental standards" (p. 41).

72. I am grateful to Paul R. Portney, president, Resources for the Future, and to Robert N. Stavins, Albert Pratt Professor of Business and Government at Harvard's John F. Kennedy School for Government, for helping me understand economists' analyses of these issues. Neither is responsible for remaining gaps in my knowledge.

73. Barry G. Rabe, "Power to the States: The Promise and Pitfalls of Decentralization," in Vig and Kraft, eds., *Environmental Policy in the 1990s*, pp. 32–33; Thad L. Beyle, *State Government* (Congressional Quarterly Press, 1997), pp. 77–81; Rivlin, *Reviving the American Dream*, pp. 102–07.

74. Rabe, "Power to the States," pp. 34–40; Enterprise for the Environment, *The Environmental Protection System in Transition: Toward a More Deisrable Future* (Washington, D.C.: Center for Strategic and International Studies, 1997), pp. 42–43; Council of State Governments, p. 127; Environmental Council of the States background information sheets; personal communications with several heads or former heads of state environmental agencies, including Mary Gade (Illinois), Kathy Prosser (Indiana), Harold Reheis (Georgia), and Robert E. Roberts (South Dakota; executive director of ECOS).

75. Davies and Mazurek, *Pollution Control*, pp. 39–48.

76. J. Clarence Davies and Jan Mazurek, *Regulating Pollution: Does the U.S. System Work?* (Resources for the Future, 1997), p. 11.

77. Steve Lohr, "Suiting Up for America's High-Tech Future," *New York Times*, December 3, 1995, sec. 3, p. 1; Martha H. Peak, "Hypergrowth in the Desert," *Management Review*, American Management Association, October 1995; Associated Press, "Intel Has Conflicts in New Mexico," June 15, 1994; Dan McGraw, "Protesting Progress," *U.S. News & World Report*, June 13, 1994, p. 60; Martha Groves, "The Cutting Edge; Land of Disenchantment," *Los Angeles Times*, May 4, 1994, pt. D, p.1.

78. John H. Cushman Jr., "States Neglecting Pollution Rules," *New York Times*, December 15, 1996, sec. 1, p. 1.

79. "Common Man with a Truly Uncommon Chance to Serve the People" (text of Governor Gilmore's inaugural address), *Washington Post*, January 18, 1998, p. A16.

80. U.S. Congress, General Accounting Office, *Water Pollution: Differences among the States in Issuing Permits Limiting the Discharge of Pollutants* (January 23, 1996), p. 15.

81. The final report of the Ozone Transport Assessment Group can be found at www.epa.gov/ttn/otag/; information about EPA's action can be found at www.epa.gov/oar/.

82. Council on Environmental Quality, *Environmental Quality*, p. 378; Margaret Kriz, "What a Waste," *National Journal*, April 6, 1996, p. 763.

83. By 1996 more than 40 states had provided tax credits or concessions for new businesses, despite doubts about their effectiveness in attracting business from out of the state. Chi and Leatherby, *State Business Incentives*, pp. 5–8. States also compete for international business. California, Michigan, Washington, and sev-

eral other states have aggressively pursued foreign markets for products and services from their state, in part as a cushion against future domestic recessions. Michael M. Phillips, "States Rely on Exports to Cushion Economic Downturn," *Wall Street Journal*, December 26, 1996, p. A2. Montana now competes with Switzerland for big foreign bank accounts with a new state law that provides unusual privacy for deposits. Timothy Egan, "Montana Has Mountains like Switzerland and Wants Numbered Accounts," *New York Times*, January 18, 1998, sec. 1, p. 16.

84. The term "liability" is used here not in the sense of a legal obligation but in the sense of an onerous undertaking.

85. For one discussion of factors influencing state approaches to environmental protection and a summary of the literature, see William R. Lowry, *The Dimensions of Federalism: State Governments and Pollution Control Policies* (Duke University Press, 1992), pp. 8–9.

86. EPA, "Compliance Assurance Implementation Plan for Concentrated Animal Feeding Operations" (U.S. EPA, 1998).

87. Council of State Governments, *ECOS*, July/August 1997, p. 1; Terry M. Neal and David Montgomery, "In Md., a 'Smart Growth' Consensus," *Washington Post*, April 5, 1997, p. A1.

88. Robert Hanley, "New Jersey Tax Rise Sought for Transit and Land Gains," *New York Times*, May 22, 1998, p. A25; James Bennet, "Land Is on the Ballot," *New York Times*, August 2, 1998, sec. 4, p. 2.

89. Margaret Kriz, "The Politics of Sprawl," *National Journal*, February 6, 1999, p. 332; Michael Janofsky, "Gore Offers Plan to Control Suburban Sprawl," *New York Times*, January 12, 1999, p. A16.

90. *Budget of the United States Government, Fiscal Year 2000* (1999), pp. 189–90.

91. Rabe, "Power to the States," pp. 44–45; Percival and others, *Environmental Regulation, pp.* 818–19.

92. Percival and others, *Environmental Regulation*, pp. 88–94.

93. Pollution abatement and control, for example, cost the United States $121.8 billion in 1994, or nearly 1.8 percent of the gross domestic product (GDP), and increased twice as fast as the GDP. (That does not include spending for conservation.) Less than 2 percent of those costs were for regulation and monitoring by government. Most of the direct cost of reducing pollution (more than 65 percent) is paid by business. The government pays 27 percent in pollution abatement activities related to sewage treatment, public utilities, trash disposal, and the like; and individuals pay the rest. Vogan, "Pollution Abatement and Control Expenditures," pp. 48–52.

94. Unfunded Mandates Reform Act, P.L. 104-4 (1995); Safe Drinking Water Amendments, 1996, P.L. 104-182.

95. U.S. General Accounting Office, "EPA and the States," GAO/RCED-95-64 (1995), pp. 16–32.

96. Beyle, *State Government*, pp. 21–22; Alan A. Altshuler and José A. Gómez-Ibáñez, *Regulation for Revenue: the Political Economy of Land Use Exactions* (Brookings, 1993), p. 63 (on the increasing costs of infrastructure due to environmental laws). After the enactment of the Resources Conservation Recovery Act in 1976, which required companies to carry liability insurance to cover

unforeseen clean-up costs, the number of hazardous waste facilities decreased by nearly two-thirds. Percival and others, *Environmental Regulation*, p. 1254.

97. Rivlin, *Reviving the American Dream*, p. 129.

98. Dana Milbank, "For Republican Governors, Spending Isn't a Dirty Word," *Wall Street Journal*, February 17, 1998, p. A24; Arturo Perez, *State Fiscal Outlook for 1997* (National Conference of State Legislatures, 1997), pp. 1–3.

99. Beyle, *State Government*, pp. xiii–xiv.

100. Beyle, *State Government*, p. 5.

101. Paul R. Portney, president of Resources for the Future, presentation at National Issues Forum, Brookings, October 13, 1998.

102. Vivian S. Toy, "Sealing Mount Garbage," *New York Times*, December 21, 1997, sec. 1, p. 41; State of Maryland, *Blue Ribbon Citizens Pfiesteria Action Commission*, Final Report, November 1997, Harry R. Hughes, Chairman, p. 12 (on nitrogen in Chesapeake Bay from air pollution); Davies and Mazurek, *Pollution Control*, pp. 16–17 ("Some 80 to 90 percent of the PCBs in the [Great Lakes] is deposited by the atmosphere.")

103. Rabe, "Power to the States," pp. 40–41.

104. Bank of America, "Economic Growth and the Environment," *Economic and Business Outlook*, June/July 1993, p. 2.

105. Thomas Michael Power, *Lost Landscapes and Failed Economies: The Search for a Value of Place* (Washington, D.C.: Island Press, 1996), pp. 14, 112, 237.

106. Rivlin, *Reviving the American Dream*, pp. 134–35.

107. Allen R. Myerson, "A New Breed of Wildcatter for the 90's," *New York Times*, November 30, 1997, sec. 3, p. 1.

108. Tom Kenworthy, "In the 'New West,' Political Landscape Resists Change," *Washington Post*, October 15, 1996, p. A1.

109. Michael M. Phillips, "Tourism's Role Rises, Creating Some Risks," *Wall Street Journal*, October 7, 1996, p. A1. Tourism accounted for nearly 10 percent of all U.S. jobs in 1995, compared with about 7 percent in 1975.

110. Michael Barone and Grant Ujifusa, *The Almanac of American Politics, 1998* (National Journal, 1997), pp. 1127–28.

111. John R. Ehrenfeld and Jennifer Howard, "Setting Environmental Goals: The View from Industry," *Linking Science and Technology to Society's Environmental Goals*, a National Research Council Report (National Academy Press, 1996), pp. 281–325; Rivlin, *Reviving the American Dream*, p. 22; Vogan, "Pollution Abatement and Control Expenditures," p. 50.

112. William D. Ruckelshaus, "Stopping the Pendulum," *Environmental Forum*, Environmental Law Institute, November/December 1995, p. 26.

113. Claudia H. Deutsch, "Cooling Down the Heated Talk," *New York Times*, May 27, 1997, p. D1.

114. Warren Brown and Martha M. Hamilton, "Super-Clean Cars Can Be Ready in '98," *Washington Post*, December 18, 1997, p. A1.

115. Neil Ulman, "Timber Industry Turns to Former Opponents to Clean Up Its Act," *Wall Street Journal*, March 12, 1997, p. A1.

116. Martha M. Hamilton and Warren Brown, "Automakers Plan Low-Emission Cars," *Washington Post*, February 5, 1998, p. E1.

117. Martha M. Hamilton, "Firms Warm Up to Climate Treaty," *Washington Post*, November 2, 1998, p. A7.

118. Bruce Guile and Jared Cohon, "Sorting Out a Service-Based Economy," in Marian R. Chertow and Daniel C. Esty, eds., *Thinking Ecologically* (Yale University Press, 1997), pp. 84–85. The Aspen Institute is working with representatives of corporations and investment banks to improve the way financial markets value environmental improvements, www.aspeninst.org.

119. Thomas Petzinger Jr., "Business Achieves Greatest Efficiencies When at Its Greenest," *Wall Street Journal*, July 11, 1997, p B1.

120. Neil Ulman, "A Maine Forest Firm Prospers by Earning Eco-Friendly Label," *Wall Street Journal*, November 26, 1997, p. A1. See also the Forest Stewardship Council web page: http://www.fscus.org.

121. Michael M. Phillips, "Tourism's Role Rises, Creating Some Risks," *Wall Street Journal*, October 7, 1996, p. A1.

122. "The Defence of Nature: How to Be a Green Rancher," *Economist*, April 12, 1997, p. 25.

123. Greg Jaffe, "South's Growth Rate Hits Speed Bump," *Wall Street Journal*, January 15, 1997, p. A2.

124. Personal communication, James A. Schriner, vice-president of Fantus Consulting, 1996. Robert Tannenwald, "State Regulatory Policy and Economic Development," *New England Economic Review*, March/April 1997, p. 84. ("Workers may be willing to sacrifice monetary compensation in order to . . . live in communities relatively free of pollution and such areas may also attract retirees and tourists with considerable purchasing power.")

125. Jesse H. Ausubel, "The Environment for Future Business," *Pollution Prevention Review*, Winter 1998, pp. 39–52.

126. OECD, *Integrating Environment and the Economy*, p. 9. "It often makes economic sense for . . . multi-nationals to adhere to a single (high) standard in *all* their operations world-wide." Frances Cairncross, "Save a Pound and Save the Planet," *Economist* (U.K. edition), September 8, 1990, p. 6.

127. Council on Environmental Quality, *Environmental Quality*, pp. 43–44.

128. Enterprise for the Environment, *The Environmental Protection System in Transition*, p. 56.

129. Stephan Schmidheing and Bradford Gentry, "Privately Financed Sustainable Development," in Chertow and Esty, eds., *Thinking Ecologically*, p. 129.

130. Percival and others, *Environmental Regulation*, 1995 supplement, pp. 50–51; Small Business Regulatory Efficiency and Fairness Act of 1996, P.L. 104–21.

Chapter 4

1. Ronald Outen, "Environmental Pollution Laws and the Architecture of Tobacco Road," in National Research Council, *Multimedia Approaches to Pollution Control: Symposium Proceedings* (1987), p. 139.

2. U.S. Environmental Protection Agency (EPA), *The Quality of Our Nation's Water: 1994*, p. 30.

3. In this context, the spectrum ranging from aggressive to limited policies refers to the relative force behind efforts to overcome political obstacles. Obstacles might include public and private costs, resistance to particular kinds of measures such as taxes or emissions trading, opposition by powerful interest groups, or changes in everyday habits. More aggressive policies do not necessarily mean more prescriptive policies. Some policies that are arguably less prescriptive (for example, taxes and emissions trading) may at times encounter more political resistance. Nor does this continuum reveal anything about the relative cost-effectiveness of policies.

4. The time frame that today's voters will accept for future benefits of present public investments is a subject of debate among economists and political scientists, heightened in the late 1990s by the issue of global climate change. Analyzing discount rates, economists often argue that the present value of payoffs more than 25 or 30 years in the future is very small. Political scientists analyze interests that influence the time frame of particular decisions. For economists' perspectives, see Thomas C. Schelling, "The Cost of Combating Global Warming: Facing the Tradeoffs," *Foreign Affairs*, December 1997, p. 8; Richard N. Cooper, "Toward a Real Global Warming Treaty," *Foreign Affairs*, March/April 1998, p. 66. For one analysis of political factors, see John Donahue, *These Disunited States*, pp. 63–66.

5. Executive Office of the President's Council on Environmental Quality, *Environmental Quality: The Twenty-fifth Anniversary Report of the Council on Environmental Quality* (GPO, 1997), pp. 42–45, 129–48; Interagency Ecosystem Management Task Force, *The Ecosystem Approach: Healthy Ecosystems and Sustainable Economies*, 3 vols. (U.S. Department of Commerce, 1995).

6. EPA, *Clean Water Action Plan Overview*, February 14, 1998; U.S. Congress, General Accounting Office, *Water Pollution: Differences among the States in Issuing Permits Limiting the Discharge of Pollutants*, January, 1996, p. 15.

7. Congressional Budget Office, *Federalism and Environmental Protection: Case Studies for Drinking Water and Ground-Level Ozone* (November 1997), pp. 41–46; John J. Fialka, "EPA Rules Seek 28% Cut in Emissions Blamed for Smog, Rankling Some Utilities," *Wall Street Journal*, September 25, 1998, p. B2; John H. Cushman Jr., "U.S. Orders Cleaner Air in 22 States," *New York Times*, September 25, 1998, p. A14.

8. President's Council on Sustainable Development, *Sustainable America: A New Consensus for the Future* (February 1996), chap. 2, available at www.whitehouse.gov/pcsd/publications. National Academy of Public Administration, *Setting Priorities, Getting Results—A New Direction for EPA* (April 1995), pp. 81–98; Enterprise for the Environment, *The Environmental Protection System in Transition: Toward a More Desirable Future, 1997*, p. 5; Council on Environmental Quality, *Environmental Quality*, p. 31.

9. Robert V. Percival and others, *Environmental Regulation: Law, Science, and Policy* (Boston: Little, Brown, 1995 supplement), p. 821.

10. Peter J. Howe, "MWRA Board Votes Down Filtration," *Boston Globe*, October 22, 1998, p. B1.

11. Council on Environmental Quality, *Environmental Quality*, p. 236; U.S. Congress, General Accounting Office, *EPA and the States: Environmental Challenges Require a Better Working Relationship* (April 1995), p. 18; Congressional Budget Office, *Federalism and Environmental Protection*, pp. 17–33.

12. Congressional Budget Office, *Federalism and Environmental Protection*, p. 37.

13. Congressional Budget Office, *Federalism and Environmental Protection*, pp. 50–63.

14. Council on Environmental Quality, *Environmental Quality*, pp. 139–40; James Brooke, "Land Trusts Multiplying, Study Shows," *Washington Post*, October 1, 1998, p. A20; Dan Eggen, "Preservation Groups Sweeping the Land; More Nonprofits Protect More Acreage," *Washington Post*, October 1, 1998, p. D4.

15. Council on Environmental Quality, *Environmental Quality*, pp. 24–25.

16. *Budget of the United States Government, Fiscal Year 1999*, pp. 77–80; Council on Environmental Quality, *Environmental Quality*, p. 43.

17. Margaret Kriz, "Feuding with the Feds," *National Journal*, August 9, 1997, pp. 1598–1601.

18. In January 1998 Massachusetts's plan to let 10,000 dry cleaners and photo processors self-certify annual compliance with air pollution rules degenerated into a dispute with federal officials, who threatened to penalize the state unless it required businesses to keep records for five years, instead of a state-proposed requirement of three years, according to the state's commissioner. "Mass. Nears Deal with USEPA to Expand Permit Alternatives," *State Environmental Monitor*, January 12, 1998, pp. 4–5.

19. Percival and others, *Environmental Regulation*, p.989.

20. Percival and others, *Environmental Regulation*, p. 714 (they discuss the fact that negotiation can be more time-consuming than litigation). See also Enterprise for the Environment, "The Environmental Protection System in Transition: Toward a More Desirable Future" (1997), p. 50 (which discusses the fact that negotiations can be extremely expensive and slow, can consume large amounts of executive time in travel and meetings, and can get bogged down by entrenched interests or burnout); Cary Coglianese, "Assessing Consensus: The Promise and Performance of Negotiated Rulemaking," *Duke Law Journal*, vol. 46 (April 1997), p. 1255 (which suggests that "reg neg" may not improve speed or reduce litigation); Peter Schuck, "Legal Complexity: Some Causes, Consequences, and Cures," *Duke Law Journal* vol. 42 (1992), p. 1 (discussing the costs of complex policies).

21. *Budget of the United States Government, Fiscal Year 1999*, pp. 77, 80.

22. Science Advisory Board, "Integrated Environmental Policy-Making in the 21st Century" (EPA draft, 1999), chap. 6.

23. Council on Environmental Quality, *Environmental Quality*, p. 287.

24. Percival and others, *Environmental Regulation*, p. 151; J. Clarence Davies and Jan Mazurek, *Pollution Control in the United States: Evaluating the System* (Resources for the Future, 1998), p. 211.

25. Tom Tietenberg, *Environmental and Natural Resource Economics*, 3d ed. (HarperCollins, 1992), pp. 164–70.

26. Davies and Mazurek, *Pollution Control in the United States*, p. 140; National Research Council, *Linking Science and Technology to Society's Environmental Goals* (National Academy Press, 1996), p. 30. EPA's trading proposal for nitrogen oxides is described at www.epa.gov/oar/.

27. Illinois EPA, 1997 report.

28. David M. Halbfinger, "New Tax Incentive for Natural Gas Trucks Could Mean Cleaner Air for the South Bronx," *New York Times,* January 18, 1998, p. A27.

29. Council on Environmental Quality, *Environmental Quality*, p. 357.

30. "The Defence of Nature: How to Be a Green Rancher," *Economist*, April 12, 1997, p. 25.

31. Percival and others, *Environmental Regulation,* p. 439.

32. Enterprise for the Environment, "The Environmental Protection System in Transition," pp. 37–38 (parking subsidies); *Budget of the United States Government, Fiscal Year 1999,* p. 179 (timber and mining tax provisions).

33. Science Advisory Board, "Integrated Environmental Policy-Making in the 21st Century," chap. 6; Council on Environmental Quality, *Environmental Quality*, pp. 38–40.

34. The idea of secrecy as government regulation has been explored extensively by Senator Daniel Patrick Moynihan in *Secrecy* (Yale University Press, 1998).

35. Safe Drinking Water Act Amendments of 1996, P.L.104-182. The requirement exempted systems serving fewer than 10,000 customers.

36. National Academy of Public Administration, *Resolving the Paradox of Environmental Protection—An Agenda for Congress, EPA, and the States*, September 1997, p. 33.

37. *Budget of the United States Government, Fiscal Year 1999,* p. 79; the Environmental Defense Fund's Scorecard project posts information about toxic releases by zip code at http://www.scorecard.org.

38. Ross Kerber, "When Is a Satellite Photo an Unreasonable Search?" *Wall Street Journal,* January 27, 1998, p. B1.

39. "Three Midwestern States Boost Public Access to Industrial Data," *State Environmental Monitor,* March 2, 1998, p. 15; John H. Cushman, Jr., "E.P.A. Is Pressing Plan to Publicize Pollution Data," *New York Times*, August 12, 1997, p. A1; "Sun Co. Shines in a Report Card on Oil Companies' Environmental Performance," *Wall Street Journal*, February 12, 1998, p. A1.

40. "Idaho Employs Novel Method in Developing Water Pollution Caps," *State Environmental Monitor*, January 12, 1998, p. 5; Marie R. Chertow and Daniel C. Esty, *Thinking Ecologically: The Next Generation of Environmental Policy* (Yale University Press, 1997), p. 177.

41. See, for example, General Accounting Office, *Environmental Enforcement: EPA Cannot Ensure the Accuracy of Self-Reported Compliance Monitoring Data*, GAO/ RCED-93-21 (GAO, March 1993), and *Water Pollution: Many Violations Have Not Received Appropriate Enforcement Attention* (GAO, March 1996). See also Clifford S. Russell, "Monitoring and Enforcement," in Paul R. Portney, ed., *Public Policies for Environmental Protection* (Resources for the Future, 1990), pp. 249–56.

42. "Utility-Owned Land: Can It Be Preserved Even As It Is Developed?" *New York Times,* January 18, 1999, special section on utilities.

43. Sevine Ercmann, "Enforcement of Environmental Law in United States and European Law: Realities and Expectations," *Environmental Law Writer* (1996), pp. 1214–38.

44. Percival and others, *Environmental Regulation,* pp. 132–33.

45. John H. Cushman Jr., "Solution Announced to an Obscure Environmental Problem: The Television," *New York Times,* January 9, 1998, p. A12.

46. Wayne R. Ott and John W. Roberts, "Everyday Exposure to Toxic Pollutants," *Scientific American,* February 1998, p. 91.

47. Louis Uchitelle, "The Economy Grows. The Smokestacks Shrink.," *New York Times,* November 29, 1998, Business, p. 4.

48. Tibbett L. Speer, "Growing the Green Market," *American Demographics,* August 1997, p. 45–49.

49. Neil Ulman, "Going Green: A Main Forest Firm Prospers by Earning Eco-Friendly Label—Seven Islands Garners Sale While Relying on Birds to Curb the Budworms—But Inspections Rile Critics," *Wall Street Journal,* November 26, 1997, p. A1.

50. Science Advisory Board, "Integrated Environmental Decision-Making in the 21st Century, chap. 6.

51. Peter Hoffman, "Going Organic, Clumsily," *New York Times,* March 24, 1998, p. A23.

Epilogue

1. Stephen S. Hall, *Mapping the Next Millennium* (Vintage, 1993), relates for general readers some of the advances in knowledge about the earth. Robert Kates and William C. Clark, "Expecting the Unexpected?", *Environment,* March 1996, p. 6, explain the inevitable limits of such knowledge and the importance of improving decisions made under conditions of uncertainty. J. Clarence Davies and Jan Mazurek, *Pollution Control in the United States: Evaluating the System* (Resources for the Future, 1998), pp. 54–96 (quote at p. 96), describe government efforts to improve information and explain in detail gaps in critical information. The National Research Council, *Linking Science and Technology to Society's Environmental Goals* (National Academy Press, 1996) examines the need for better information to support policy decisions.

2. During the 1990s, the National Research Council issued several detailed reports assessing progress and shortcomings in environmental research, with emphasis on improvements needed to provide an effective underpinning for policy. Reports include National Research Council, *Research to Protect, Restore, and Manage the Environment* (National Academy Press, 1993); *A Biological Survey for the Nation* (National Academy Press, 1993); and *Linking Science and Technology to Society's Environmental Goals* (National Academy Press, 1996).

3. Charles Mann and Mark Plummer, "Qualified Thumbs Up for Habitat Plan Science: Assessment of Habitat Conservation Plans," *Science,* December 19, 1997, p. 2052.

4. National Academy of Engineering, *Keeping Pace with Science and Engineering: Case Studies in Environmental Regulation* (National Academy Press, 1993), pp. 101–03.

5. Beth Baker, "How Science Faired under the 104th Congress," *Bioscience,* January 1997, p. 10.

6. National Research Council, *Linking Science and Technology to Society's Environmental Goals* (National Academy Press, 1996), pp. 41–42.

7. Information about the Conservancy's Natural Heritage programs' databases is located at http://www.tnc.org; personal communication, Douglas Hall, vice-president of the Nature Conservancy, 1997.

8. Mary Graham, "High Resolution, Unresolved," *Atlantic Monthly,* July 1996, pp. 24–28; John J. Fialka, "Fears of Terrorism, Chemicals Clash in EPA Debate," *Wall Street Journal,* September 3, 1998, p. A24.

9. U.S. Department of Commerce, "Pollution Abatement and Control"; J. Clarence Davies and Jan Mazurek, *Pollution Control in the United States: Evaluating the System* (Resources for the Future, 1998), pp. 286–87.

10. U.S. Department of Commerce, "Pollution Abatement and Control"; J. Clarence Davies, *Pollution Control,* pp. 286–87; Albert H. Teich, "The Federal Budget and Environmental Priorities," in National Research Council, *Linking Science and Technology to Society's Environmental Goals,* pp. 365–66.

11. See, for example, Enterprise for the Environment, *The Environmental Protection System,* pp. 19–24.

Index

Acid rain, 67, 68, 77, 103
Administrative Procedures Act, 100
Agriculture. *See* Farming
Air pollution: atmospheric, 19; coarse and fine particulates, 19–20, 36, 56; contribution of nitrogen, 15, 20, 80; governmental responsibilities, 91–92; hazardous, 20, 119*n*2; health risks of, 19–20; indoor pollution, 6, 8, 18–19, 25, 108; interstate and regional issues, 68, 91–92, 94–95; motor vehicle, 20–21, 54, 56, 61, 85; nonattainment areas, 98; reduction in pollutants, 6, 19; research, 127*n*40; resistance to legislative measures, 25; sources of, 18–21, 38, 98; standards, 97. *See also* Automobile industry; Clean Air Act; Pollution
Alliance for Environmental Innovation, 62, 131*n*36

Ambrose, Stephen, 53
American Forest and Paper Association, 84
American Honda Motor Company, 8
Arizona, 106
Army Corps of Engineers, 26, 36
Aspen Institute, The, 137*n*117
Audits, environmental, 107–08
Audubon Society, 39, 63
Ausubel, Jesse H., 86
Automobile industry: current pollution threats, 5, 19, 20, 21; driving and commuting habits, 61, 117; fines against, 8; fuel-economy standards, 65; gasoline taxes, 103; incentives, 104; new cars, 21; oil embargo, 53–54; pollution control devices and technology, 54, 61; pollution control rules and, 2, 48, 98; pollution and emission standards, 33, 44, 50, 56, 84, 85;